MAPS

A Historical Survey of Their Study and Collecting

Published for the Hermon Dunlap Smith Center
for the History of Cartography
The Newberry Library

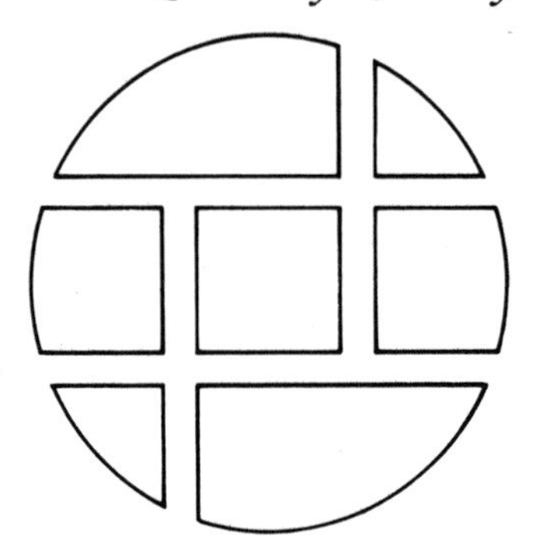

MAPS

A Historical Survey of Their Study and Collecting

R. A. SKELTON

The Kenneth Nebenzahl, Jr., Lectures in the History of Cartography at The Newberry Library

THE UNIVERSITY OF CHICAGO PRESS
CHICAGO AND LONDON

The University of Chicago Press, Chicago 60637
The University of Chicago Press, Ltd., London

Printed in the United States of America

International Standard Book Number: 0-226-76164-9
Library of Congress Catalog Card Number: 72-650049

In Memoriam R. A. Skelton

{1906-1970}

Contents

Introduction

When Mr. and Mrs. Kenneth Nebenzahl decided to endow a triennial series of lectures some nine years ago in honor of their son, Kenneth Nebenzahl, Jr. (1954–1971), they did so because of their great love for historical maps and because of the appropriateness of such a program to the Newberry Library. The trustees were delighted to accept their generous gift, viewing it as a way of increasing the proper use of the library and of stimulating research in this important field of the Newberry's holdings.

Our collections of maps and atlases have long been considered among the greatest strengths of the library. This fact can be traced back to Edward E. Ayer's interest in the discovery and exploration of the Americas, an interest that figured prominently in the forming of his great collection devoted to American Indians and their interaction with Western civilization. The Ayer Collection, given to the library in 1911, regularly augmented by gifts from Mr. Ayer until his death in 1927 and since then from his endowment, has such highlights as the Henry Stevens-Ayer Collection of editions of Ptolemy's *Geography* (including four manuscript versions), eight portolan atlases, thirteen portolan charts, and a very good collection of general maps and atlases and geographical works published before 1800.

In addition to the Edward E. Ayer cartographic materials, the Newberry has many more maps and atlases in its General Collection and in special collections, notably Everett D. Graff's on the American West and William B. Greenlee's world-famous gathering of Portuguese materials before 1830. All this forms an important corpus of material for the serious study of the history of cartography. It was to encourage that study that the Nebenzahls established their lecture series.

The establishment of the Nebenzahl lectures was announced on 3 June 1964, at a public lecture entitled "The Life and Works of Abraham Ortelius" by the distinguished Professor Cornelis Koeman of the Geografisch Instituut der Rijksuniversiteit, Utrecht. A brilliant exhibit of Ortelius's works, drawn from the Ayer Collection, had been mounted for the occasion, and a celebration dinner was held afterwards, hosted by the library's benefactors. There, serious plans began to be laid to select the inaugural Nebenzahl lecturer.

The first Kenneth Nebenzahl, Jr., Lectures in the History of Cartography at the Newberry Library took place on 27 and 28 October and on 10 and 11 November 1966. As the dean of the history of cartography in the English-speaking world, R. A. Skelton—better known to his many friends and colleagues throughout the world as Peter Skelton—was the obvious choice to give them. This first series was entitled "The Study and Collecting of Early Maps: A Historical Survey." Then Superintendent of the Map Room of the British Museum, Dr. Skelton's knowledge of the field was at once broad and deep, and he was familiar with, and often an expert in, the many subsidiary disciplines touched by the history of cartography, to which he was dedicated. A glance at the bibliography of his published works (see this volume, pp. 111–31) will illustrate this fact.

While at the Newberry, Dr. Skelton assisted in the preparation of a notable exhibition of maps from the library's collections, assembled in honor of the new lecture series, and made a survey of our holdings that has proved invaluable in formulating a policy for strengthening them further. It was his knowledge of the collections, for example, that enabled him to recommend to us one of the most notable acquisitions since his visit here: a collection of sixteenth-century maps printed in Italy, assembled by Signor Franco Novacco of Venice as part of his collection of maps, atlases, and geographical works. Of the part of his collection that the library acquired, Dr. Skelton wrote: "the Novacco Collection stands alone and is unsurpassed even by the oldest national libraries."

Further, it was Peter Skelton who first brought David Woodward to the attention of Mr. Nebenzahl, and thus to the Library. Mr. Woodward, now our Curator of Maps, was then completing his Ph.D. degree in geography at the University of Wisconsin, initiated on a Fulbright travel grant. Dr. Skelton helped persuade him to accept our offer of a fellowship and then of a permanent appointment.

Finally, it was Dr. Skelton who organized a seminar at the Newberry in 1966 to explore fully our desire to create a Center for the History of Cartography, just now come to fruition. He strongly seconded our belief in the need for such a center, and in the validity of establishing it at the Newberry, to provide an impetus for research and publication in this undeveloped and neglected discipline. In the spring of 1971 such a Center was established, by means of the gift of an endowed special fund for that purpose and with the cooperation of eleven midwestern universities through their agent, the Committee on Institutional Cooperation. It is hoped that the new center will attract support to build further the Newberry's map collections, to make them more usable to scholars not only of the history of cartography but of many other disciplines as well, and to encourage the collection and study of maps as scholarly documents of art, science, and technology.

Dr. Skelton retired as Superintendent of the Map Room of the British Museum in 1967 and immediately embarked on a long list of scholarly projects for which he had not hitherto had the time as an administrator and librarian, although he was a remarkably productive scholar. Many of the projects, alas, were still uncompleted when he lost his life following an automobile accident in December 1970. The essays in this book represent one such project. As published now, they have been edited and annotated by David Woodward and Robert Karrow of the Newberry staff. The trustees and staff of the Newberry take particular pleasure in seeing the first series of lectures in print.

These essays help to fill a significant gap in the literature of the history of cartography. By describing the methods of study and the collecting of maps rather than the maps themselves, they fall perhaps more into the field of historiography than of cartography. The first essay, a brief summary of the history of cartography itself from the Middle Ages to the twentieth century, provides a useful background. The second essay describes the many factors that have contributed to the disappearance of maps from the historical record. Many maps that survive owe a great debt to collectors who have salvaged them over the years and to archives and libraries that have preserved them. Dr. Skelton highlights the major persons and institutions that have played such a part in history. In the third and fourth essay, respectively, Dr. Skelton outlines the historical study of early maps in the past and in the present, illustrating how the history of cartography has developed as a field of study. In closing, he enumerates some of the more important projects which lie ahead for the researcher and a tentative program of ten tasks to make early maps more available and to provide more sophisticated methods for their study. Some of these tasks were in Peter Skelton's own program of projected work in his retirement: it is very much to be hoped that this book will inspire others to continue the work.

HERMON DUNLAP SMITH, *President*
The Newberry Library

Editor's Note

Owing to Dr. Skelton's tragic death in 1970, the manuscript for this book came to us unfinished. Had the author been able to see it through press, I am sure that there would have been many emendations and amplifications. In some of the more obvious editorial matters, I have tried to act for the author in a manner that I trust would have met with his approval, but the general concept and shape of his lectures remains completely unchanged as he delivered them.

The bibliographical references were compiled from Dr. Skelton's incomplete notes, which consisted only of the last names of the authors. Explanatory notes have occasionally been added to passages of the text. A few references will also be found to works published since the lectures were given in 1966, when such works superseded earlier research or added substantially to it.

Those who attended Dr. Skelton's first lecture will recollect that it was illustrated with many slides of maps highlighting the history of cartography since the Middle Ages. As many of these maps have already been reproduced in Dr. Skelton's *Decorative Printed Maps of the 15th to 18th Centuries* (London: Staples Press, 1952), in Leo Bagrow's *History of Cartography* (London: C. A. Watts, 1964), and in other readily available general works, it seemed needless to include them here.

I should like to acknowledge the generous help I have received from many individuals in editing these lectures. Drs. George Kish, Arthur H. Robinson, and Helen Wallis have kindly read through the manuscript and provided helpful comments. I am particularly grateful to Mr. Robert W. Karrow, Jr., Map Cataloger at the Newberry Library, who spent many hard hours tracing and checking Dr. Skelton's references in numerous libraries, and who compiled the bibliography of the author's works at the end of this volume with the kind help of Mrs. Alexa Barrow, Mr. John Huddy, Mrs. Mary Skelton, and Dr. Helen Wallis.

In the final essay it will be noticed that Dr. Skelton appeared neither satisfied with the status quo nor very encouraged for the future. Since the lectures were delivered, there have been many encouraging signs that the field is at last coming into its own: for example, in carto-bibliographical tools alone, we now have *The British Museum Catalogue of Printed Maps, Charts, and Plans* (London: Trustees of The British Museum, 1967) in fifteen volumes; Professor Cornelis Koeman's great *Atlantes Neerlandici* (Amsterdam: Theatrum Orbis Terrarum, 1967–71) in five volumes; James Clement Wheat and Christian Brun's *Maps and Charts Published in America Before 1800* (New Haven, 1969); the *Dictionary Catalog of the Map Division of the New York Public Library* (Boston: G. K. Hall, 1971) in ten volumes; and the National Maritime Museum's *Catalogue of the Library* (London: H.M.S.O., 1971), of which volume 3 (parts 1 and 2) is devoted to atlases and cartography. The achievements in facsimile publishing in the field, emanating particularly from the Netherlands, have likewise been encouraging. The international conferences in the history of cartography, of which the first two were held at London in 1964 and 1967, continued their success at Brussels in 1969 and at Edinburgh in 1971. The vitality of these biennial meetings of international scholars in the field bodes well for the future. Were Dr. Skelton able to complete this manuscript, I am sure that he would have incorporated this information (and much more) into his essays, or added it as a postscript.

There is, however, one aspect of current work that, through his own modesty, he would not have emphasized or perhaps even have mentioned: his own writing. The bibliography of Dr. Skelton's published works demonstrates his magnificent contribution to the literature of the history of cartography, and his own great role in creating the growing interest in it.

David Woodward

MAPS

A Historical Survey of Their Study and Collecting

R. A. Skelton

1

The History of Cartography: An Introductory Survey

MAPS HAVE MANY FUNCTIONS AND MANY FACES, AND each of us sees them with different eyes. Every map, of whatever date or purpose, is a synthesis of experience, theoretical concepts, and technical craftsmanship. In viewing it, some will look for evidence of historical facts; others for the reflection of the mapmaker's knowledge, thought, and state of mind; others again for the formal qualities of design which distinguish the map as an artefact or a work of art.

Maps are graphic documents of specialized purpose and inbred type, and the diversity of personal approach to them should not blind us to the essential continuity of the historical process by which they have evolved and are still evolving. Even modern maps yield more readily to analysis and classification if they are considered the end products of a continuous evolutionary process, reaching back into the Middle Ages, in which common elements are expressed in developing forms and with increasing precision and widening application.

Visual symbols are more tenacious of life than verbal modes of expression. In Roman itineraries and in modern subway maps we find similar conventions of design used for kindred purposes. The topographical map of today, though constructed by sophisticated techniques of measurement and computation, employs a vocabulary of expression which is largely not of recent origin. Some of its symbols were introduced in the fifteenth and sixteenth centuries, others in the eighteenth or early nineteenth; its coloring conventions certainly go back to the Middle Ages and probably to Roman cartography. Even the graphic design of thematic or "special-purpose" maps—usually regarded as a relatively recent innovation—has its roots in the first half of the nineteenth century.

"History develops, Art stands still." This phrase, coined by E. M. Forster, is relevant to the story of mapmaking and of the collection and study of early maps. Every map, in the strict sense of the word, contains a kernel of geographical or topographical fact which dictates the basic form of visual presentation and places limits on the freedom or fantasy of its maker. Basically the map is a linear design elaborated by pictorial or geometrical symbols and accompanied by geographical names and legends. During the last seven centuries the mapreader has had to adapt himself to change far less than the bookreader.

But this seemingly simple design is the outcome of complex processes—assembly of information, both graphic and textual; assimilation to the mapreader's geographical ideas, to transmitted cartographic patterns, or to political interest; and the practical stages of compilation, control, adjustment, and copying. Study of this background illuminates the mapmaker's working methods; it enables us to look over his shoulder and perceive why he drew an outline or attached a place name to a feature; and it throws light on the pressures to which he was subject.

While the "look of maps" has transformed itself slowly, the content of maps has undergone continuous change through time. It is this changing content that gives maps significance as documents for social, economic, and political history. Since geography may often condition men's lives, maps are necessary tools for students in these fields, and they furnish information which no other kind of document can give so efficiently. In turn, new information and new kinds of experience tend to evoke new and changing forms of expression in mapmaking, as in other graphic arts. In this sense, any map is a precipitation of the spirit and practice of its time. Its design and decoration may be of interest to the art historian, to the student of calligraphy and typography, even to the social historian and the psychologist. Its construction, reflecting advances in techniques of survey and representation, is a matter for the historian of science and technology. The physical structure of atlases or books of maps is a proper, and often perplexing, study for the bibliographer.

The development of mapmaking as a craft has not followed a smooth upward curve. It is marked by points of rapidly accelerated advance, followed by periods of standstill or even retrogression. Throughout the history of cartography the forces of inertia (the "built-in image") and of change are exerted in a constantly varying relationship. In evaluating an early map, the student should carefully balance the respective effect of these forces and should never underrate the weight of inertia.

The history of the collecting and study of maps, to which the later lectures are devoted, cannot be dissociated from that of map production. I propose to trace briefly the evolution of mapmaking as a craft during the period, since the Dark Ages, from which specimens have survived into our day. The mental picture which it suggests is that of a track leading over successive false crests or watersheds, each higher than the last, but separated by sections of level or downhill going. Without taking a cataclysmic view, we may find it helpful to isolate in our minds the points of more prominent relief. They give us a series of chronological horizons as an index to the competence and character of mapmaking in any given era, and they dictate the critical controls which should govern the study of maps.

In cosmography and mapmaking the medieval church was the heir of the Roman Empire, and adopted the modes of Roman thought. Until the tenth century the writings and maps of Latin geographers of the first to the fifth centuries provided the Christian mappamundi with its pattern of land and sea and with 90 percent of its nomenclature. The production centers of monastic cartography lay in those provinces of the empire which had been most thoroughly Romanized. Although the series of surviving world maps is interrupted, there was no gap in transmission; and in this field of thought the Middle Ages form a bridge between two creative periods in the description of the earth—that of the Greeks and that of the Renaissance. The medieval world map was drawn from literary sources; it was schematic in form and didactic in intention; and only very slowly, in the fourteenth and fifteenth centuries, was it to admit information from experience.

The regional and local maps which made their appearance in Western Europe in the thirteenth century represent an altogether different kind of development. These maps were statements of geographical fact, prepared from observation or measurement and designed for the practical use (on the one hand) of travelers by land or sea and (on the other) of tenants of settled land. They cannot be traced back to any Roman or Byzantine models, even if these existed. It is from the second half of the thirteenth century that we have the earliest surviving nautical charts and the earliest post-Roman road maps and local surveys.

The most primitive of these maps reflect the verbal descriptions from which they evolved. It is to the thirteenth century—the age of Albertus Magnus and Roger Bacon—that we ascribe this first great turning point in modern cartographic history: the recognition that a graphic design will communicate geographical relationships more efficiently than a written document, and that the elements of a map can only be transmitted by text or by word of mouth with great difficulty.

The topographical map was, in comparison with the sea chart, slower to free itself from its written models. Medieval land surveying was a regular process in the settlement of customary rights and dues and in determining the economic relationship between landlord and tenant; but very few of the numerous surveys of this kind made before 1500 are now accompanied by maps. The road maps drawn by Matthew Paris about 1250 are little more than illustrated itineraries or lists of road stations with distances; and his maps of England are adapted to the itinerary form.[1] The tenacity of life of the "itinerary map" is illustrated by the reappearance of the "strip map" form, not only in John Ogilby's road atlas of England published in 1675,[2] but also in the route maps issued today by motoring organizations. The "Gough" road map of Great Britain, drawn in the early fourteenth century, is exceptional, for the Middle Ages, in its correctness of outline and proportion, and King Edward I's surveyors, on whose measurements the map was apparently based, reached a degree of accomplishment in their craft unsurpassed in the next two hundred years.[3]

The first written Italian *portolano*, or set of sailing directions, dates from about 1250, and within the next half-century the prototype of the Mediterranean sailing chart seems to have come into existence. The original model then established was reproduced with no structural alterations for nearly four centuries—a fact which argues successful adaptation of form to function. The earliest known chart, the so-called Carte

1. The Matthew Paris maps are reproduced in *Four Maps of Great Britain Designed by Matthew Paris about A.D. 1250* (London: Trustees of the British Museum, 1928).

2. For a discussion of Ogilby's road atlas, see Herbert G. Fordham, "John Ogilby (1600–1676), His *Britannia* and the British Itineraries of the Eighteenth Century," *The Library*, 4th ser., 6 (1925–26): 157–78.

3. E. J. S. Parsons' *The Map of Great Britain circa A.D. 1360 Known as the Gough Map: An Introduction to the Facsimile . . . with the Roads of the Gough Map by Sir Frank Stenton* (Oxford: University Press, 1958) is accompanied by an excellent collotype reproduction of the map in color. See also R. A. Pelham, "The Gough Map," *Geographical Journal* 81 (1933): 34–39.

Pisane, was drawn (it is supposed) a little before 1300. It shows us how the normal portolan chart was constructed. First, the scaffolding: the central windrose and its rays, the subsidiary windroses on the periphery of a circle centered on the main rose. Then the geographical outlines: in transferring them to the vellum the windrays are used as a guide, and—outside the circle of rays—a grid of squared lines is drawn for this purpose. Lastly, the nomenclature and decoration.

The portolan chart was developed in the Italian maritime republics, where the mercantile middle class was becoming familiar with mensuration and mathematics. Charts of this type were constructed, and used, by angular measurement, which was unknown in land surveying until the sixteenth century. They served the seaborne commerce of the Italian and Catalan ports and embraced its trade routes from the Black Sea to Flanders.

The demand was plain. From 1270 onward, records tell of charts carried in ships. In the fourteenth century, chartmakers' shops were active in the seaports of North Italy and Catalonia. Men who had learned their trade by drawing pilot charts of the Mediterranean drew the world maps of the fourteenth and fifteenth centuries which incorporated into the traditional mappamundi new data from experience; and they made some marked improvements in its design. It was the chartmakers of southern Europe, too, who were to map the oceanic discoveries, in the East and West, made during the fifteenth and early sixteenth centuries.

Until the middle of the fifteenth century, the chartmakers were (so far as we know) the only active professional cartographers; and even they often combined this trade with that of instrument maker or shipmaster or astrologer. The recovery, translation, and copying of Ptolemy's *Geographia*, with the maps found in the Byzantine manuscripts, were to call into existence in Italy the professional mapmaker, a man competent in the mathematical cartography to which Ptolemy introduced the West. This new kind of craftsman grew up alongside the professional printer. The printing press not only gave much wider currency to his work but also enabled it to be reproduced more faithfully.

Here is the second significant watershed of history. It is a stage in which the construction of maps was transformed by the use of mathematical projections and grids; in which the content of maps was immensely enriched by oceanic discovery and more accurate instrumental observation; in which mapmaking became an independent profession; and in which geographical consciousness at all levels was awakened by the printed map.

The cartographers who began their practice in the fifteenth century were, like the first printers, largely recruited from pictorial artists—painters, miniaturists, and illuminators. These men, who were not necessarily, or by original vocation, geographers, copied the Latin manuscripts of Ptolemy, and were engaged in the copying, decoration, and—ultimately—compilation of maps. In the "modern" maps which they added to the manuscripts and printed editions of Ptolemy, they extended his world picture to the north and west to take in Scandinavia and Greenland.

The maps of Henricus Martellus, drawn at Florence about 1490, form a bridge between Ptolemy's world picture and that of the Great Discoveries, and illustrate the common inspiration of the search for eastern and western seaways to East Asia—the one pioneered by the Portuguese, the other by Columbus. Martellus worked with Francesco Rosselli, the first specialized map-printer and dealer known to us, whose workshop in Florence has a particular significance in the diffusion of new geographical information. Rosselli seems to have been interested in the mathematical basis of his art, for he was associated with three new map projections devised to accommodate the expanding world picture of his time.[4]

Martellus's maps point back as well as forward. In amending Ptolemy's map of Asia he took into account the Portuguese entry into the Indian Ocean in 1488, Marco Polo's description of the Far East, and Toscanelli's views on the longitudinal extension of Eurasia and the width of the western ocean. The resulting representation of the Old World, and particularly of Asia, became the traditional base map upon which the discoveries in the east and west were to be grafted by cartographers between 1492 and 1510. This was the world map that came under the eyes of Columbus and Cabot, of Martin Behaim and Waldseemüller, and of the chartmakers of Lisbon and the Italian cities.[5] It con-

4. On Rosselli see Roberto Almagià, "On the Cartographic Work of Francesco Rosselli," *Imago Mundi* 8 (1951): 27–34, and George E. Nunn, *World Map of Francesco Roselli* [sic] *Drawn on an Oval Projection and Printed from a Woodcut Supplementing the Fifteenth Century Maps in the Second Edition of the Isolario of Bartolomeo dali Sonetti* (Philadelphia, privately printed, 1928).

5. Roberto Almagià, "I mappamondi di Enrico Martello e alcuni concetti geografici di Cristoforo Colombo," *La Bibliofilia* 42 (1940): 288–311.

veyed a quantitative statement of the width of ocean supposed by its author to separate Europe from East Asia; this agreed closely with the estimates of Columbus and Toscanelli and with Columbus's expectations and identifications of landfalls.

The wide diffusion of this prototype, the most mature version of which is represented by the wall map at Yale,[6] suggests that it was engraved, and illustrates the influence of a cartographic pattern to which the printing press had given currency. Here—as in other fields of science and technology—engraving and printing revolutionized the communication of visual information. They introduced "the exactly repeatable pictorial statement."[7]

Before printing, when multiplication was the work of hand copyists, the repetition of visual statements was subject to infinite variation and corruption, and the promulgation of the data which they alone could convey was either very imperfect or not even attempted. In geography, as in botany and other sciences of observation, the absence of reliable graphic illustration robbed the verbal statement of definition. The door was closed to comparative study and classification. The first great advance in comparative cartog-

6. For a description of the Yale map, see the section by R. A. Skelton, "Mappemonde de Henricus Martellus Germanus, New Haven (c. 1490)," in Marcel Destombes, ed., *Mappemondes, A.D. 1200–1500*, Monumenta cartographica vetustioris aevi, vol. 1, Imago Mundi Supplements, vol. 4 (Amsterdam: N. Israel, 1964), pp. 229–34, and Alexander O. Vietor, "A Pre-Columbian Map of the World, circa 1489," *Imago Mundi* 17 (1963): 95–96.

7. William M. Ivins, Jr., *Prints and Visual Communication* (Cambridge: Harvard University Press, 1953), p. 2.

raphy was therefore the application or printing to map reproduction.

The printing press also made the cartographer independent of the "V.I.P." patron. It enabled him to put his work on sale in the street and the market place; and Rosselli's obvious commercial prosperity is in strong contrast to the meager financial harvest reaped by the great Italian artist-craftsmen of his day. The inventory of Rosselli's stock made in 1527 contains the earliest Italian record of the rolling press used for printing from copperplates.[8] In comparison with the older screw press, the rolling press was capable of more rapid output and greater uniformity of impression. It was essentially an instrument for mass production. Here we see the technological basis of a cartographic industry producing, selling, and exporting identical impressions of the maps drawn in the master's workshop.

Rosselli's business may be considered as specializing in cartographic work—he was a Rand McNally or Bartholomew of his day. But map materials also reached, and were reproduced by, the journeymen-printers who turned out the best-selling literature of tracts and pamphlets, in vernacular languages as well as Latin, by which the news of the discoveries was spread. These little works—the paperbacks of the six-

8. "Inventory of the Contents of the Shop of Alessandro di Francesco Rosselli," in Arthur M. Hind, *Early Italian Engraving: A Critical Catalogue with Complete Reproduction of All the Prints Described. Part 1. Florentine Engravings and Anonymous Prints of Other Schools*, 4 vols. (New York: M. Knoedler, 1938), 1:304–9. The significance of this inventory for printing history is discussed in Henry Meier, "The Origin of the Printing and Roller Press," *Print Collector's Quarterly* 28 (1941): 189–90.

teenth century—carried maps in tens of thousands into almost every country in Europe.

In the cartographic activity of the two crucial decades 1490–1510 we can discern in embryo much of the development of cartography during the following century. Many elements, already foreshadowed, were to be elaborated with far greater technical resources. New map projections were devised, mainly to give a more faithful representation of high latitudes. The extent and relationship of the continents and the possible existence of navigable passages through or around them were anxiously studied and conjecturally laid down on the maps. The woodcut was superseded by the copperplate as a more effective medium for map printing, and an immensely productive map trade grew up, first in Italy and later in the Netherlands, commercially organized to supply the European market. The gradual accretion of new maps to augment the twenty-seven Ptolemaic maps anticipated the appearance of the atlas, the typical product of synoptic cartography, as an index to contemporary knowledge of the world.

In mapping and cartography three new elements—two technical, one political—characterize the sixteenth century and make of it a third watershed. The usefulness of the chart as a tool for navigation was enhanced by study of the problems arising from the variation of the compass and from the convergence of the meridians. Mercator first drew rhumb lines correctly as spirals on his globe of 1541; and on the world chart of 1569, drawn on his new projection, the seaman could for the first time lay down a compass course as a straight line. As the mathematical basis of navigation became better understood, a pilot could plot his landfalls with greater accuracy and more hope of recovering them. Secondly, the introduction of geometrical methods of land survey—in particular, triangulation—and the development of new and more precise instruments enabled the surveyor to cover the ground more quickly, to observe horizontal angles, and to ascertain distances by indirect measurement. He could now, for the first time, determine the distance between two points without walking over the ground between them. His work gained both in speed and in precision, and with these technical advances he could, and did, undertake the mapping of counties, provinces, and countries. Regional cartography gathered momentum, especially in Central Europe.

This century saw also the growth and entrenchment of the national state in Europe. Political and commercial rivalries were sharpened; ambitions for territorial expansion and for new markets clashed; and the questions raised by contemplation of the world map—the continuity of the New World, its relationship to East Asia, the possibility of northern passages, the existence of a Southern Continent—became geopolitical issues.

The technical revolution in topographical survey was also harnessed to the national interest. It is generally true that the mapmaker's resources, in tools and methods, have developed most rapidly in periods of accelerated economic and social change. In Tudor England the surveyor's technical advance had an essentially agrarian background; and his profession emerged as a new trade or "mystery" with a functional part to play in the life of the country, both social and political. In 1570–79 Christopher Saxton, an estate surveyor, mapped the counties of England and Wales; and in 1577 he obtained a license to sell the engraved maps. But long before this, early proofs of them were being delivered to Lord Burghley, who annotated them in his fine hand with details of political and military intelligence.[9]

In other countries—Spain, Portugal, the Netherlands—the national state exercised its patronage and control over cartography. During the sixteenth and seventeenth centuries, governments showed increasing awareness of the significance of maps. In the political field, maps served for the demarcation of frontiers; in the economic, for property assessment and taxation, and (eventually) as an inventory of national resources; in administration, for communications; in military affairs, for both strategic and tactical planning, offensive and defensive.

From the middle of the seventeenth century, these trends became accentuated. The productive cartographic industries of the Netherlands, France, Germany, and England continued to exercise a conservative influence. The life of a copperplate was prolonged, by husbandry and reworking, long after the map engraved on it had outlived its reliability; for the commercial map dealer the "sale of the work" (in Captain Cook's words) came first. But original mapping emerged as essentially an instrument of national policy, partly conscious and officially controlled,

9. See Edward Lynam's introduction to the facsimile edition: *An Atlas of England and Wales: The maps of Christopher Saxton, Engraved 1574–1579* (London: Trustees of the British Museum, 1936).

partly spontaneous and inspired by private enterprise. This is the threshold of the period (extending into our own day) in which scientific inquiry and technological practice have been pressed into the service of the state. It begins the age of the government scientist. The great exploratory surveys of the eighteenth and early nineteenth centuries–in the Pacific Ocean, in Africa, in the Arctic, and in continental North America–were, almost without exception, prompted by political or economic motives, even if the elements of scientific research became increasingly conspicuous in them. The art of the military surveyor was perfected in France and Germany, and he was responsible for the topographical mapping of many countries before the establishment of national survey departments. It was a military engineer, Captain Samuel Holland, who, on the beach at Louisbourg on the day after its surrender in July 1758, introduced James Cook to the use of surveying instruments and began his instruction in trigonometry and astronomy.[10] And in the mid-eighteenth century there were better maps of some parts of the American colonies, made by military engineers, than there were of the English Home Counties. In the United States, the creation of mapping agencies by President Jefferson and his successors was foreshadowed by the military surveys initiated by Washington in the Revolutionary War.[11]

The so-called "reformation of cartography" which

10. R. A. Skelton, "Captain James Cook as a Hydrographer," *Mariner's Mirror* 40 (1954): 92–119.

11. Herman R. Friis, "A Brief Review of the Development and Status of Geographical and Cartographical Activities of the United States Government: 1776–1818," *Imago Mundi* 19 (1965): 68–80.

took place, mainly on French initiative, between about 1670 and 1750, laid the geodetic foundations both for national mapping and for the world map, and provided the basis for eighteenth-century cartography.[12] It involved (in succession) measurement of arcs of the meridian, to ascertain the size and figure of the earth; astronomical observations to determine accurately the position of a great number of places on the earth's surface, in latitude and longitude; and survey of large areas, by triangulation from precisely measured bases and with improved instruments. The number of positions accurately fixed increased from 40 in 1682 to 109 in 1706 (and by 1817 to over 6,000). Between 1733 and 1783 the whole of France was covered by a network of triangulation, which became the basis for the celebrated Cassini map.

In England the government was reluctant to undertake the expense of surveying the country and was at length pressed into doing so by the enterprise of the military engineer William Roy. When he died in 1790, General Roy left a specification for "a general survey of the British Islands"; and a year later the Duke of Richmond, as Master-General of the Ordnance, created a regular establishment for the Trigonometrical Survey under the auspices of the Board of Ordnance (which was of course a military department of government).[13]

12. Christian Sandler, *Die Reformation der Kartographie um 1700* (Munich: Oldenbourg, 1905) and R. A. Skelton, "Cartography," in *A History of Technology*, 5 vols., ed. Charles Singer (Oxford: University Press, 1954–58), 4:596–628.

13. R. A. Skelton, "The Origins of the Ordnance Survey of Great Britain," *Geographical Journal* 128 (1962): 415–26.

The new standards of precision to which these first national surveys were made depended on the technological accomplishment of instrument makers, particularly those of London, who supplied the European market in the eighteenth century. English craftsmen perfected the theodolite, the basic tool of topographic survey; and the "great theodolites," three feet in diameter, constructed by Jesse Ramsden for the surveys of Roy and the Board of Ordnance, measured azimuthal angles at seventy miles to an accuracy of two seconds of arc. The successive development of the bubble level and the *Y*-level and of the aneroid barometer enabled absolute altitude, from a common datum, to be determined; this was something new, and of vital significance both to the economic projector and to the scientist. The new leveling instruments made possible the canal building of the eighteenth century and the railway building of the nineteenth. Barometric determination of height enabled Alexander von Humboldt and the German geographers who followed him to correlate altitude with the distribution or vertical zones of natural phenomena.

The mapmakers incorporated this new material by the adoption of two technical conventions. These were the spot height, first used in England by Christopher Packe in 1743, and the contour line.[14]

14. François de Dainville, "De la profondeur à l'altitude; des origines marines de l'expression cartographique du relief terrestre par cotes et courbes de niveau," in *Le navire et l'économie maritime du moyen-âge au XVIIIe siècle principalement en Méditerranée*, ed. Michel Mollat (Paris: S.E.V.P.E.N., 1958): 195–213. This article was reprinted in the *International Yearbook of Cartography* 2 (1962): 150–62, and has been translated into English by Arthur H. Robinson:

The contour is a special application of one of the cartographer's basic forms of expression—the isometric line. Its roots are in the representation of the depth of the seabed (isobaths). An early printed example is found on the map of the Merwede estuary by Cruquius (1729). The idea was adapted to represent land elevations at the end of the eighteenth century, and by the second quarter of the nineteenth the contour had become a standard cartographic device, gradually superseding the hachuring of the late eighteenth and early nineteenth centuries.

It is in the eighteenth century that the continuous enrichment of the content of maps really begins—a process which has continued with progressive acceleration into our own day. In step with it, cartographers have developed increasingly flexible modes of expression and representation, both in topographical and thematic mapping. This was the mapmaker's response to a double challenge: in the economic field, and in the scientific.

In the economic field, changes in the tenure and use of land—in the eighteenth as in the sixteenth century—increased its yield and so its value. The English instrument maker George Adams, in his *Geometrical and Graphical Essays* (London, 1791), pointed to the surveyor's responsibility for accurate measurement of so valuable a commodity as land; nothing less than the best instruments should be used, although (he added) in America, where land was much cheaper, the sur-

---

"From the Depths to the Heights: Concerning the Maritime Origins of the Cartographic Expression of Terrestrial Relief by Numbers and Contour Lines," *Surveying and Mapping* 30 (1970): 389–403. See also Arthur H. Robinson, "The Genealogy of the Isopleth," *Cartographic Journal* 8 (1971): 49–53.

veyor could get by with a mere plane table. In Europe, the agricultural improvements of the eighteenth century associated with new crops and farming techniques called for more precise physiographic delineation of the land than the topographical mapmakers could offer. This demand evoked some ingenious mapping of soils and land use, foreshadowing methods in use today. Parallel with these essays in the delineation of the surface of the land and its cover, the young science of geology developed new representational techniques for mapping the underlying rocks, for classifying their formations, and for studying the sculpture of land forms. French cartographers of the eighteenth century made some attempts to summarize geological and mineralogical observations over large areas; but not until William Smith, "mineral surveyor," made his momentous correlation of rocks with their associated fossils, at the end of the century, did the modern geological map emerge.

In other earth sciences too—meteorology, hydrology, the geography of plants, animals, and man—data obtained from observation and measurement were becoming abundant. Toward the middle of the nineteenth century, current knowledge of physical geography could be presented synoptically, as we can see in the several editions of the remarkable *Physikalischer Atlas* of Heinrich Berghaus and Keith Johnston.[15]

It was not only in the field of natural science that more penetrating and more exact observations, in much greater quantity, gave the mapmakers novel tasks in

15. Gerhard Engelmann, "Der Physikalische Atlas des Heinrich Berghaus; die kartographische Technik der ältesten thematischen Kartensammlung," *International Yearbook of Cartography* 4 (1964): 154–61.

representation. In social science also the same process was at work. The facts of human geography—population, settlement, occupation, communications, and other aspects of the work, wealth, and life of men in their environment—were being recorded by statistical or actuarial techniques; the first national census of the United States was taken in 1790, and the first of Great Britain in 1801. Maps were brought into service to depict and collate these data in quantitative form.

The cartographers of the eighteenth century, particularly in France and England, had introduced into their work a great variety of detail relating to social life and economic activity; and their maps are a precious source for the historian. But the vocabulary of signs which they used—although ingenious and much richer than anything found in earlier medium-scale maps—was not different in kind from the naïve pictorial symbolism of the sixteenth-century map. To show the data of the physical and social sciences in geographical relationship, the nineteenth-century cartographer needed an extra dimension for his maps. This he found in the use of tone and color.

Maps had of course been colored from the earliest times, mainly to distinguish political divisions or physical features (water or woodland), or for decorative purposes. Apart from a few isolated experiments in two- or three-color printing of maps, all color was added to the map by hand until the middle of the nineteenth century.[16] This was in fact still being done in the large-scale plans of the Ordnance Survey and in the maps of

16. R. A. Skelton, "Colour in Mapmaking," *Geographical Magazine* 32 (1960): 544–53.

the Geological Survey of Great Britain until about 1900. But the introduction of color lithography about 1840 triggered a technical revolution which was to drive out the use of handwork and aquatint and profoundly influenced map production. It placed at the printer's disposal a mechanical process for reproducing flat colors; it eliminated the great labor of applying color separately to each copy and of checking the correctness of each copy; and it ensured uniformity throughout the printing of a map. If we add to this the consideration that the cartographer's linear design could be drawn directly on or transferred mechanically to the lithographic stone or zinc plate, it is clear that the risk of corruption by the engraver was removed or much reduced. This was a technical advance not less significant than the first printing of maps from copperplates or woodblocks in the fifteenth century. With it the mapmaker's medium for reproduction became infinitely more flexible and expressive.[17]

Even before the middle of the nineteenth century, that is, before the advent of chromolithography, mapmakers had anticipated some of the processes used today for depicting the facts of human geography. It has been said that in the seminal period 1835–55 "almost every technique now known for representing population numbers, distribution, density and movements seems to have come into being."[18] A pioneer in this was the English engineer officer H. D. Harness,

17. Kenneth Nebenzahl, "A Stone Thrown at the Mapmaker." *Papers of the Bibliographical Society of America* 55 (1961): 283–88.

18. Arthur H. Robinson, "The 1837 Maps of Henry Drury Harness," *Geographical Journal* 121 (1955): 440.

who prepared maps to illustrate the report of a railway commission for Ireland published in 1838. In his traffic maps Harness drew "streams of shade" proportionate in thickness to the movements represented; this is the earliest use of what are now called "flow lines." In the population map Harness indicated towns and cities by solid black circles proportionate in size to the population figures, and the density of rural population was shown by four degrees of aquatint shading.

The stage of cartographic history which set in during the early nineteenth century cannot properly be described as a watershed. We seem rather to come over the lip of a plateau which rises in a steady upward gradient to the middle of the twentieth century. The materials for the world map were immensely enriched by exploratory survey using light precision instruments such as the chronometer and the reflecting sextant (which George Adams called "a portable observatory"). International standards for constructing the world map had begun to appear. The meter, adopted in 1791 in France, provided a natural linear measure related to the meridian circumference of the earth; the representative fraction, first used in France in 1806, is an international language of scale. These are among the foundations of the International 1:1,000,000 Map of the World, for which specifications were agreed to by thirty-five countries in 1913.

From 1814 the Hydrographic Office of the Admiralty set in motion its program for extending British naval surveys to the coasts and waters of almost the whole world. Exploration of the floor of the ocean, a "sealed volume" until deep-sea sounding could be undertaken, began also. M. F. Maury's chart of the North

Atlantic (1853) gave over a hundred soundings of depths greater than 2,000 fathoms (about two miles); and Maury showed the way to systematic charting of ocean currents and winds.

In topographical mapping, between 1750 and 1850, trigonometrical surveys were initiated by twenty-one European states, besides India. The thematic map had before 1850 established itself as a vital tool for scientists in many fields, working in depth, recording their data quantitatively or statistically, and requiring a plotting of their findings in geographical relationship or distribution. The area in which maps serve society has expanded continuously from that period, as the frontiers of knowledge are extended and the materials to be used in maps multiply.

It is possible now to perceive a mid-twentieth-century watershed in cartographic history, based on the dramatic development of air survey, photogrammetry, and a number of electronic distance-measuring devices. In the field of map reproduction, significant improvements in quality have been attained by the use of plastic drawing media and more sophisticated offset-lithography techniques.

The genetic evolution of the map has been continuous from the Middle Ages to the present day. The materials for reconstructing it, to the end of the sixteenth century and even later, are sadly defective and fragmentary. To the history of cartography, perhaps more than to any other historical study, we must apply La Rochefoucauld's dictum: "History never embraces more than a small part of reality."

2
The Preservation and Collecting of Early Maps

It is safe to say that the wastage or loss of early maps, up to the sixteenth century and even later, has been more severe than that of any other class of historical document. Generalizations founded by historians of cartography on surviving examples must often be taken with a pinch of salt. Many links in the chain of transmission are lost. The discovery of a hitherto unknown map may, like the turning of a kaleidoscope, recast the accepted pattern of thought and hypothesis or provide a "missing link" whose existence had been conjectured.

Suppose that the literary historian of today knew only one poem of Chaucer; the historian of printing, no works from the presses of Gutenberg and Schoeffer; the musicologist, no versions of the folk tunes used by Haydn and Holst? The map historian is in just such a case; and if he is prudent he will concede the imperfect or provisional nature of the constructions which he builds on all too slender foundations. The imprudent map historian of course gets more fun out of his excursions to the lunatic fringe—and even beyond it; but he wastes a good deal of our time.

We owe the greater debt to those collectors who—

like Sir Robert Cotton, "the worthy repairer of eating time's ruins"[1]—have salvaged so many early maps as we still have for our study and enjoyment. T. S. Eliot has described tradition as a "means by which the vitality of the past enriches the life of the present," and the map is a vehicle for tradition. Few map collectors of the past were indiscriminate accumulators. Their motives were enumerated by Dr. John Dee, in his preface to the *English Euclid* (1570):

> Some to beautify their Halls, Parlers, Chambers, Galeries, Studies, or Libraries with; other some, for things past, as battles fought, earthquakes, heavenly firings, and such occurrences, in histories mentioned: thereby lively as it were to view the place, the region adjoining, the distance from us, and such other circumstances: some other, presently to view the large dominion of the Turk: the wide Empire of the Muscovite: and the little morcel of ground where Christiandom (by profession) is certainly known, little I say in respect of the rest, etc.: some other for their own journeys directing into far lands, or to understand other men's travels . . . liketh, loveth, getteth and useth, Maps, Charts, and Geographical Globes.[2]

Here Dee distinguishes four main reasons for con-

1. This tribute to Cotton appears in the unpaged "Summary Conclusion of the Whole," in vol. 2 of John Speed's *History of Great Britaine under the Conquests of ye Romans, Saxons, Danes and Normans* (London: John Sudbury & Georg Humble, 1611).

2. *The Elements of Geometrie of the Most Auncient Philosopher Evclide of Megara*, trans. H. Billingsley (London: I. Daye, 1570), fol. a4^{r}.

temporary map collecting. Reversing his order, maps are (1) aids to travel, (2) aids to contemporary studies, (3) aids to historical studies, and (4) decorative objects.

Maps have had, and still have, a low survival rate. This is particularly true of sheet maps, and notably of the great wall maps which in the late Middle Ages and the Renaissance provided a synthesis of contemporary knowledge. There have been many factors which have contributed to the scarcity of early maps. First we have to remember that a high proportion of early maps were not made to last. A chartmaker or land surveyor of the sixteenth or seventeenth century (for instance) kept his eyes on the present and had no idea that he might be creating a historical document. He was simply doing a professional job for his employer or customer; and it is this very naïveté which makes his work useful as a record in the perspective of history. But the motive for creation was not necessarily a motive for preservation. There is a dangerous interval of vulnerability between the moment at which the practical usefulness of a map is exhausted and the moment at which it awakes the interest of historians as a relic or memorial of the past.

The length of this time lag, in respect of any particular map, depends partly on the rate of change in the geographical or human situation represented in the map, and partly on the development of attitudes toward the past. In 1795 Joseph Lindley, who had just surveyed the county of Surrey, wrote in the memoir published with his map: "To the antiquarians of the 25th century I hope the map will be useful as affording them an opportunity of determining the positions of many places in the County of which the devouring hand of time may not, perhaps, have left the last vestige remaining."[3] The devouring hand of time, and of other agents, is doubtless as familiar in Illinois as in Surrey. But Lindley's belief that his map would have to wait *six centuries* before being useful to the local historian indicates in a revealing way the rate of change to which men had been accustomed before the nineteenth century and by which they measured that of the future. The vulnerable interval during which a map might be discarded or destroyed as no longer of practical utility and not yet of historical interest was then much greater, and likely to be bridged only by some accident of survival.

Maps tend to be superseded by others more modern in content, or presented in a form better adapted to their purpose and use, or executed by a new technique. This process may be called the second enemy of survival. The compilation materials which we know to have been incorporated in many Renaissance maps are lost, and we can only guess at their form and content from the use made of them in the derivatives. In designing his globe of 1492, Martin Behaim had at Nuremberg a large printed world map by, or after, Henricus Martellus Germanus[4] and—for his rendering of the Arctic—the fourteenth-century tract known as

3. Joseph Lindley and William Crosley, *Memoir of a Map of the County of Surrey, from a Survey Made in the Years 1789 and 1790* (London: G. Bigg, 1793).

4. On Behaim's debt to Martellus, see Armando Cortesão, *Cartografia e cartógrafos Portugueses dos séculos XV e XVI*, 2 vols. (Lisbon: Edição da "Seara nova," 1935), 1: 126–35, 140–42.

Inventio Fortunata. Neither of these is now extant, although both seem to have come under the eyes of not a few contemporaries, including Christopher Columbus. The Martellus map is probably to be identified with one of the large world maps of which the plates or blocks were listed in the stocks of Alessandro di Francesco Rosselli dispersed in or after 1527; no example of these great printed maps is known, and if they then went out of circulation it was because their world picture had been rendered obsolete by the great discoveries which expanded it. The *Inventio*, which was apparently never printed, could be taken seriously as a source by Mercator as late as 1569; but eight years later the copy he used had been lost to sight, and our only knowledge of this widely diffused text comes from the abstract of it which Mercator wrote for John Dee in 1577.[5]

The extension of knowledge may have discredited older maps and caused them to be discarded or destroyed. Not less potent a factor in this wastage has been the introduction of new cartographic forms and techniques. Even if some princely collectors of the fifteenth century, such as Federigo Duke of Urbino, still preferred manuscript to printed books,[6] there can

5. E. G. R. Taylor, "A Letter Dated 1577 from Mercator to John Dee," *Imago Mundi* 13 (1956): 56-68, and Tryggvi J. Oleson, *Early Voyages and Northern Approaches* (Toronto: Canadian Centenary Series, 1968), pp. 105–8.

6. "In this library all the books are superlatively good, and written with the pen, and had there been one printed volume it would have been ashamed in such company." *Renaissance Princes, Popes, and Prelates: The Vespasiano Memoirs, Lives of Illustrious Men of the XVth Century*, trans. William George and Emily Waters (New York: Harper & Row, 1963), p. 104.

be no doubt that from the sixteenth century the printed map prevailed and that many owners jettisoned manuscript maps when engraved or woodcut ones became available. Is this surprising if we consider the fact that each and every hand copy was liable to corruption in transcription, and the consequences of such errors were (for instance) a matter of life and death on a nautical chart used by seamen for navigation? Similarly, when lithography and zincography were applied to map reproduction in the nineteenth century, engraved copperplates of maps were discarded or beaten down for re-use in great numbers, so that they are now extremely rare.

From the scarcity of many early atlases known to have been printed in substantial editions we can only suppose that they were thrown aside when improved methods of atlas production were introduced. Ortelius's *Theatrum orbis terrarum* (1570), the pioneer of the modern atlas, was planned to provide a conspectus of world geography in a form more convenient than the bound-up sets of Italian maps, which lacked its uniformity of authorship, scale, and format. We must wonder how many such sets of Italian maps perished when their owners purchased the *Theatrum*. The rarity of sea atlases printed in the Netherlands in the sixteenth and seventeenth centuries, in relation to the number and size of editions, doubtless has a similar explanation; out-of-date charts were not only useless, they were also dangerous. Even the early editions of important nineteenth-century atlases which were often revised and republished, such as Stieler's *Hand-Atlas* and the physical atlases of Berghaus and Johnston, are now rarely found in the market.

The third, and not least dangerous, enemy of maps is their physical form. Sheet maps normally lack the protection of a binding cover. Wall maps, by virtue of their size and exposure to varying conditions of light, temperature, and humidity, are especially vulnerable, and it is among them that mortality has been highest. Those which, in the sixteenth century, came from the printing presses of Italy, Germany, and the Netherlands survive in one or two copies or not at all; and those produced by Hondius, Blaeu, Visscher, and other Amsterdam engravers of the next century are little less scarce. In England, the tithe surveys made after 1836 are the earliest attempt at large-scale mapping of the whole country and present the most complete record of the agrarian landscape at any period; but the enormous parish maps which resulted have sadly deteriorated in so short a life span as a hundred years.[7]

Lastly, the catastrophes to which human affairs are subject—fire, earthquake, war, revolution, and political upheaval—have caused the destruction or disappearance of an incalculable quantity of early maps. While we can only guess at the extent of our loss in some classes, the identity and contents of others are known or can be inferred from catalogues and similar records: for instance, the maps in monastic libraries scattered at the Reformation; Portuguese charts drawn before 1500 (of which only two are known); the stock of the London book trade ravaged by the Great Fire of 1666, and that of Joan Blaeu burned in his printing house in 1672; the treasures of European libraries lost

7. J. B. Harley, "Maps for the Local Historian: A Guide to British Sources, 3. Enclosure and Tithe Maps," *Amateur Historian* 7 (1967): 265–74.

in the two world wars of this century. To these sinister agents we must add the uncontrolled dispersal or alienation of special collections built up with discrimination and expertise. In 1780 Richard Gough wrote that "many capital collections of MSS have been dispersed irrevocably"; and the process has continued into our own day, to the grievous loss of systematic scholarship. Mercator's library was sold at auction by his sons in 1604, and even the sale catalogue (still extant in 1915) has now disappeared. Among important cartographic records which have quite recently suffered fragmentation or extinction as an entity are those of the *Armeebibliothek* in Munich, of the Dartmouth family, of Leo Bagrow, of the English collector Cyril Kenney, and—we may add—the correspondence of Abraham Ortelius.[8]

8. Losses of important cartographic records due to the second World War were recorded in notices in *Imago Mundi* under the title "With Fire and Sword." On the Armeebibliothek collection, see vol. 10 (1953), p. 56. The Dartmouth library was dispersed by auction: Sotheby & Co., *Catalogue of Valuable Printed Books, Important Manuscript Maps, Autograph Letters and Historical Documents, etc., the Property of the Earl of Dartmouth . . . March 8 & 9, 1948* (London: Sotheby, 1948). The Bagrow collection was partially destroyed during the second World War; his collection was dispersed after his death, the Russian maps going to Harvard University: "With Fire and Sword," *Imago Mundi* 4 (1947): 30, and Mary M. Bryan, "The Harvard College Library Map Collection," Special Libraries Association, *Geography and Map Division Bulletin* 36 (1959): 7. The Kenney collection, including a great many books on early surveying and technology in addition to the maps and atlases, was sold in six portions and recorded in Sotheby & Co., *Catalogue of the Celebrated Collection the Property of C. E. Kenney* (London: Sotheby, 1965–68). Ortelius's correspondence was printed in J. H. Hessels, *Abraham Ortelii et virorum erudi-*

What then are the conditions for survival of early maps, or (put another way) to what circumstances do we owe the preservation of those which have come down to us?

We can distinguish between two modes of transmission. Some maps, more particularly those created in the course of official business, have remained in their place of origin among public or private muniments; the archives of a nation, a city state, or a municipality; the papers of a statesman, a soldier, or a surveyor; the records of a business house. Implicit in "organic" or "static" repositories of this character is a will to preserve—or an inertia—which at once screens their collections from normal hazards and furnishes a continuous record of their history.

Maps which have suffered migration or displacement enjoy no such guarantees. Their survival depends on one or other of three contingencies: first, possession by successive owners (not always known to us) who thought them worthy of care and preservation; second, the awakening of interest in them by students while the maps were still available; and third, chance or accident. Some maps which have reached us by such channels have a traceable story. But, as M. R. James remarked of ancient manuscripts, "very many have none. We can

---

torum ad eundem et ad Jacobum Coluam Ortelianum epistulae, Ecclesiae Londino-Batavae archivum, vol. 1 (Cambridge: University Press, 1887), and sold at Sotheby's on 14 February 1955. Thirteen years later, the collection, still almost intact (i.e., comprising some 250 letters to him and 27 by him) was resold at Sotheby's and thus scattered throughout the world: "The Correspondence of Abraham Ortelius," in *Catalogue of Valuable Continental Books and Autograph Letters . . . 17 & 18 June, 1968* (London: Sotheby & Co., 1968), pp. 81–112.

only say of them that they were written in such a century and such a country, and acquired at such a date: and there an end."[9] The world map of Juan de la Cosa was drawn (as its signature asserts) in 1500, and it turned up in a bric-à-brac shop in Paris in 1832. What happened to it between these two dates is completely unchronicled; we do not know who owned it or how it was preserved. The authenticity and date of a document may be confirmed by evidence of provenance and history—but they are not invalidated by the want of such evidence.

The story of deliberate and purposeful map collecting begins in the Renaissance, more precisely at the turn of the fifteenth and sixteenth centuries. There were several reasons for this, to which I shall return. The most significant were the growth of map production in both volume and variety, the canalization of public interest toward geography, and the development of historical or antiquarian motives. In the absence of these factors, the formation of a distinct collection of maps during the Middle Ages was inconceivable and in fact unknown.

The quantity of maps drawn in the medieval period should not be underestimated; even if we take into account only surviving examples, it was far greater than is commonly thought. An inventory published in 1964 enumerated over 1,100 manuscript *mappaemundi* executed before the beginning of the sixteenth century

9. Montague R. James, *The Wanderings and Homes of Manuscripts* (London: Society for Promoting Christian Knowledge, 1919), pp. 10–11.

and still, or recently, in existence.[10] But we must note that all but a score of these are found in codices and were drawn as illustrations to texts copied and recopied for monastic libraries. Of the larger world maps drawn before 1400, and thus of the truly medieval type betraying its Roman origins, no more than three, with fragments of two others, survived into the twentieth century.[11]

The maps of the Middle Ages exhibit a very limited number of different types. A simple scheme of classification accommodates all known *mappaemundi*, which were designed as a synthesis of familiar knowledge expressed in a well-established form and providing a vehicle for iconographic illustration of sacred and profane lore.[12] This was the principal motive for which they were made, valued, and preserved. Not a few maps were thought worthy of mention in the medieval library inventories.[13] Virtually all were world maps, in a variety of physical forms. They were described as painted, "historié et figuré," on vellum or canvas, rolled or in codex or on folding panels, preserved in cases of leather or wood, or executed on the walls of church porticoes.

10. International Geographical Union, *Monumenta cartographica vetustioris aevi A. D. 1200–1500* (Amsterdam: N. Israel, 1964—), vol. 1, *Mappemondes A. D. 1200–1500*, Imago Mundi Supplements, vol. 4, ed. Marcel Destombes.

11. Ibid., pp. 193–205.

12. Three influential schemes for classifying medieval maps are those of Destombes (ibid., pp. 15–20); Michael Corbet Andrews, "The Study and Classification of Medieval Mappae Mundi," *Archaeologia* 75 (1925–26): 61–76; and Richard Uhden, "Zur Herkunft und Systematik der mittelalterlichen Weltkarten," *Geographische Zeitschrift* 37 (1931): 321–40.

13. [Leo Bagrow], "Old Inventories of Maps," *Imago Mundi* 5 (1948): 18–20.

Among the maps so recorded, mainly in monastic libraries, were the late Roman prototypes (no longer extant) from which medieval cartographers constructed their *mappaemundi.* A fourth-century version of the world map prepared by Vipsanius Agrippa for the emperor Augustus must have been available to Richard of Haldingham, who made the Hereford *mappamundi* at the end of the thirteenth century. Most of the maps in medieval libraries however were drawn in the scriptoria of the monastic houses which preserved them, and a relatively small proportion migrated from their place of origin before the Renaissance humanists came to rummage the monastic libraries in search of ancient manuscripts. The spirit in which the holdings of medieval libraries were created and maintained is illustrated by book curses or maledictions, for example: "This book belongs to St. Maximin at his monastery of Micy, which Abbot Peter caused to be written, and with his own labour corrected and punctuated, and dedicated to God and St. Maximin . . . with this imprecation, that he who should take it thence . . . should incur damnation with the traitor Judas, with Annas, Caiaphas and Pilate, Amen."

In spite of such supernatural protection, the map production of those conventual orders which, in the later Middle Ages, did most to free geographical thought from its scholastic straitjacket has been most unevenly preserved. The achievements of the thirteenth-century Franciscans in mathematical cartography and in the reformation of the map of Asia were neglected in the next century. We have no map by Roger Bacon, by William of Rubruck, or by Marco Polo. The maps of the Benedictine Matthew Paris were salvaged by sixteenth-century collectors in England.

Fra Mauro's *mappamundi* of 1459, the crown of medieval church cartography, remained in his Camaldolese monastery at Murano until the nineteenth century. The creative map work done in the Augustinian and Benedictine houses of Southern Germany during the fifteenth century found a home, mainly after the secularization of these houses in 1803, in the court libraries of Bavaria and Austria.

In the dynamic formation of private libraries in fifteenth-century Italy, the center of the Renaissance book world, geography and cartography had a conspicuous share, even if we cannot yet point to a specific map collection. The motives which inspired the secular book collector are reflected as clearly in these subject fields as in any other: the recovery of classical learning, the spirit of inquiry into man and his environment, the sense of beauty and of glory, the passion for owning books, antiquities, and works of art as a projection of the collector's personality. It is a time when "every Italian noble forms a library," and scholars are active in copying books for their own use and enjoyment. Princely collectors employ secretaries (like Poggio Bracciolini) to ransack older libraries, and dealers (like Vespasiano da Bisticci) to procure books; they also have in their service scribes and artists regularly engaged in copying and illuminating.

The collecting of manuscripts thus "made to order" —that is, commissioned by or dedicated to a patron— extended to cartographic materials. From many possible illustrations of this fact we may select three.

1. Of the Latin version of Ptolemy's *Geographia* made by Jacopo d'Angelo, forty-one manuscripts with

maps written in the fifteenth century are known to survive; all of these were executed in Italy, and eighteen are still in Italian libraries.[14] Of these, four came into the Vatican Library founded by Pope Nicholas V in 1447, three to the great library assembled at Urbino by Federigo da Montefeltro (and eventually incorporated in that of the Vatican), and four to the library established by Cosimo de Medici in 1444.

2. Between about 1480 and 1484 Francesco Rosselli was at the court of Matthias Corvinus, king of Hungary, employed in mapmaking. The great Corvina Library, sacked by the Turks in 1526, included other cartographic works executed for the king, including charts by Venetian makers.[15]

3. Two successive rulers of the Este dynasty at Ferrara were actively interested in geography. Duke Borso d'Este (d. 1471) was a patron both of the illuminator Taddeo Crivelli, who was to promote the first printed atlas (that of Bologna 1477), and also of Donnus Nicolaus Germanus, whose versions of the Ptolemaic maps were the basis for those in almost all the early editions.[16] The reign of Borso's heir, Ercole d'Este (1471–1505), spanned the discovery of America

14. The manuscripts of the *Geographia* are enumerated and described in Joseph Fischer, *C. Ptolomaei Geographiae Codex Urbinas graecus 82*, 2 vols. in 4 (Leiden: E. J. Brill; Leipzig: O. Harrassowitz, 1932), vol. 1.

15. Florio Banfi, *Gli albori della cartografia in Ungheria; Francesco Rosselli alla corte di Mattia Corvino* (Rome: Biblioteca dell' Accademia d' Ungheria, 1947).

16. The roles of Crivelli and Nicolaus Germanus in the production of the Bologna Ptolemy (1477) are discussed in R. A. Skelton's introduction to the TOT reprint edition: Claudius Ptolomaeus, *Cosmographia, Bologna 1477* (Amsterdam: Theatrum Orbis Terrarum, 1964), pp. v–ix.

and that of the sea routes to India. In 1494 he went to Florence in search of Toscanelli's manuscripts and maps. In 1502 he obtained through his secret agent in Lisbon, Alberto Cantino, a world chart showing the Portuguese discoveries.[17] Here we come upon political and economic reasons for map collecting.

Both the making and the collection of maps were stimulated by the search for new oceanic trade routes. In the fifteenth and sixteenth centuries Venice, as the only cosmopolitan power in Italy apart from the Papacy, maintained an intelligence system and an archival organization of remarkable complexity and effectiveness. The only surviving cartographic record of Portuguese exploration before 1500 is in Venetian charts, compiled from information picked up at Cadiz or Lisbon by shipmasters on the Flanders run, or in the maps which Henricus Martellus, also using reports from the Lisbon quayside, drew at Florence.

But before 1500 Portugal had a hydrographic office charged with the custody and correction of sea charts, and a similar office was established on the same model at Seville in 1508. The political need for geographical information, and international rivalry in obtaining it, led to competition between states in acquiring not only maps but also mapmakers—as, in our own day, atomic scientists or footballers.*

17. On Cantino's relationship with d'Este, and his map, see chapter 4, "La Carte de Cantino," in Henry Harrisse, *Les Corte-Real et leurs voyages au nouveau-monde* (Paris: E. Leroux, 1883), pp. 69–158.

* Among the cartographers who, for a time at least, changed their political affiliations may be mentioned Jean Rotz, a native of France, who was appointed Hydrographer

It is for economic reasons, too, that in the period of the discoveries Franconia became a center of map collecting. The two great commercial dynasties of Augsburg, the Welsers and the Fuggers, had counters in Seville and Lisbon, and their factors sailed with the regular Portuguese fleets to Brazil and India. Back to "head office" in Augsburg or Nuremberg went Portuguese, Majorcan, or Italian maps drawn in the early decades of the sixteenth century. Many of these came into the possession of the Humanist Konrad Peutinger of Augsburg (1465–1547), who had particular opportunities for collecting in Portugal through his marriage into the Welser family and his connection with the German printer Valentim Fernandes at Lisbon. Some of Peutinger's Portuguese charts passed into the great library created by the dukes of Wolfenbüttel in the sixteenth century; but the greater part of his collections was in 1715 given to the Jesuit library in Augsburg and, on its suppression, came to the Bavarian state library.

Peutinger's interest in classical scholarship is represented by the famous Roman road map which is known by his name; and its history illustrates some of the vicissitudes to which such documents are subject. It is an eleventh- or twelfth-century copy executed in a monastic library of South Germany from a lost Roman original of the fourth century. In 1507 the Viennese scholar Konrad Celtis found the map "in a certain li-

---

to King Henry VIII and wrote an important treatise on nautical science; the Portuguese cartographers Pedro and Jorge Reinels, who enlisted in the service of Spain to prepare maps for Magellan; and André and Diogo Homem, also of Portugal, who worked, respectively, in Antwerp and London. –Ed.

brary," the identity of which he kept secret—doubtless that in which it had been made. Celtis bequeathed the map to Peutinger, who intended to publish it but never did so. After his death in 1547 it disappeared until 1591, when Marcus Welser, burgomaster of Augsburg, recovered a fragment of the map and had it engraved and published.[18] Apart from the Ptolemaic maps, this is the first example of facsimile reproduction of an early map, considered as such, by printing. Again the Peutinger Table* is lost to sight for over a century. In 1714–15 it passed by purchase in succession from Desiderius Peutinger to a book dealer and thence to Prince Eugene of Savoy, whose great library was acquired after his death in 1737 by the Vienna National Library.

The desire to own relics of classical cartography is exemplified by the diffusion of Ptolemaic manuscripts in the sixteenth century. Outside Italy, there were in this century at least six Greek and eighteen Latin manuscripts of the *Geographia*.[19] A Greek codex, bequeathed to the Dominicans of Berne in 1443 and transferred to the University Library of that city in 1539, was used by Erasmus for the first edition of the Greek text in 1533. For the editorial preparation of the Stras-

* While this translation of "Tabula Peutingeriana" is common usage now, "Peutinger Map," or "Peutinger Road Map" are better renditions of the Latin.—Ed.

18. Marcus Welser, *Fragmenta tabulae antiquae, in quîs aliquot per Rom. provincias itinera. Ex Peutingerorum bibliotheca* (Venice: apud Aldum, 1591). Seven years later, the remaining portions of the map were discovered, whereupon Welser notified Ortelius, who had the complete map engraved and published. See Christiane Piérard, "Un exemplaire de la Tabula itineraria ou Tabula Peutingeriana, édition Moretus 1598, conservé à Mons," *Quaerendo* 1 (1971): 201–16.

19. Fischer, *C. Ptolomaei Geographiae*, vol. 1.

bourg Ptolemy of 1513, Waldseemüller had attempted to borrow the Berne codex, and he succeeded in obtaining another Greek manuscript from Italy.

In Tudor England, the "discovery of British antiquity" led to the salvage of many early maps, with other historical documents, and their incorporation in private collections. The opportunity came from the breakup of medieval libraries at the dissolution of the monasteries; the motive, from the Elizabethan scholars who undertook the methodical study of the history of the British people and their institutions.[20] Thus all the surviving maps of Matthew Paris, with other examples of medieval cartography, came to rest in the libraries of Archbishop Matthew Parker and Sir Robert Cotton. Both these collections were formed in the second half of the sixteenth century, and much had already been lost, as we learn from John Leland, who surveyed monastic libraries for King Henry VIII in the years 1534–41, and from John Bale. Manuscripts were (in Bale's words) "solde to the grossers and sopesellers, and . . . sent oversee to the bokebynders, not in small nombre but at tymes whole shyppes full."[21]

Cotton was not only a zealous antiquary; he was also a man of affairs who enjoyed the power that knowledge bestowed. This was the motive for his collection of contemporary state papers for the reigns of Tudor monarchs. To this interest we owe his preservation of many maps of the sixteenth century, mainly relating

20. See Thomas D. Kendrick, *British Antiquity* (London: Methuen, 1950).

21. John Leland, *The Laboryouse Journey & Serche of John Leyland for Englandes Antiquitees, Geuen of Hym as a Newe Years Gyfte to King Henry the viii, in the xxxviii Yeare of His Reygne*, ed. John Bale (London: Johan Bale, 1549).

to political or military events. English chartwork of this century would be very imperfectly known if Cotton's acquisitive habits had not preserved it.[22] Other politicians and administrators, then as since, assembled for the practical needs of public business important map collections whose contents have survived or are recorded in early lists. Among them may be mentioned Viglius van Aytta, President of the Spanish Council of the Netherlands, whose collection (catalogued in 1575) was burned with the University Library of Louvain in 1915;[23] and William Cecil, first Lord Burghley, Lord Treasurer to Queen Elizabeth, whose maps are still preserved at Hatfield House.[24] A later English public servant also collected maps both in his private interest and in the course of duty. The library of Samuel Pepys, Secretary of the Admiralty, included many early sea charts and London plans and views of all periods; and his solicitude for his map collections is expressed in entries of his diary, for example, on 19 September 1666: "mightily troubled, and even in my sleep, at my missing . . . Speed's Chronicle and

22. R. A. Skelton, "The Hydrographic Collections of the British Museum," *Journal of the Institute of Navigation* 9 (1956): 323–34.

23. Aytta's library catalogue is printed in volume 1 of Bibliothèque Royale (Brussels), *Catalogue des manuscrits de la Bibliothèque Royale des ducs de Bourgogne*, 3 vols. (Brussels: C. Muquardt, 1839–42). See also "Collection de cartes de Viglius," *Messager des sciences historiques ou archives des arts et de la bibliographie de Belgique* (1862): 428–33.

24. On Lord Burghley's collection, see R. A. Skelton and John Summerson, *A Description of Maps and Architectural Drawings in the Collection Made by William Cecil, First Baron Burghley, Now at Hatfield House* (Oxford: Roxburghe Club, 1971), and R. A. Skelton, "The Military Surveyor's Contribution to British Cartography in the Sixteenth Century," *Imago Mundi* 24 (1970): 77–83.

Maps, and the two parts of Waggoner, and a book of cards [charts], which I suppose I have put up with too much care that I forget where they are."

By the second half of the sixteenth century the more effective organization of the European trade in books and maps encouraged the formation of private collections. The Frankfurt book fairs were regularly attended by wholesalers such as Christopher Plantin of Antwerp or Cornelius Caymox of Nuremberg. The map section in the catalogue of the 1573 fair published by Georg Willer of Augsburg included over sixty Venetian sheet maps and seventeen large wall maps mounted and painted.[25] The ledgers of Plantin record the export of maps, atlases, or globes to retail booksellers in the Netherlands, France, England, Germany, Spain, and Italy.[26]

Mapmakers and publishers were themselves collectors, and appear in Plantin's account books as in debt to him for the supply of cartographic materials; like Mercator and Ortelius, they regularly visited the Frankfurt fair; or, like Gerard de Jode, employed a factor at it. The correspondence of Ortelius illustrates the energy with which his agents in several countries sought out maps to serve as compilation materials for the *Theatrum*. But Ortelius had other motives also for collecting; his interests were predominantly historical, and his library and cabinet of antiquities at Antwerp were celebrated in his own day.

By a variant of Parkinson's Law, the immense ex-

25. See Leo Bagrow, "A Page from the History of the Distribution of Maps," *Imago Mundi* 5 (1948): 53–62.

26. Extracts from the Plantin archives relating to cartographers are printed and discussed in Jean Denucé, *Oud-nederlandsche kaartmakers in betrekking met Plantijn*, 2 vols. (Antwerp: Nederlandsche boekhandel, 1912–13).

pansion of map production during the sixteenth and seventeenth centuries—both in sheet maps and in atlas editions—gave an added impetus to collecting; and (as we shall see) the collector's needs came to be systematically served by the trade. During these two centuries, however, and even later, the independent map collection was exceptional. Although maps might be assigned to a particular class or to special presses, they were still usually administered as elements of a library or art collection. Thus the royal *Kunstkammer* and the library of Saxony, founded by the Elector Augustus (1553–86), included many maps; presumably the wall maps, mounted and framed, were in the former, and atlases and loose sheet maps in the latter. The sixteenth-century catalogues of both collections list maps among the other holdings. Not until 1701, when a fire in the palace put the collections in jeopardy, did the Elector Augustus the Strong create a separate map room. The first catalogue of this (ca.1756) recorded 117 atlases and 200 maps, making 10,000 sheets in all.[27]

The royal map collections of England provide a case history of a similar process. When the Old Royal Library was conveyed by George II to the British Museum in 1757, it contained relatively few maps and atlases. The maps listed in a catalogue made soon after the Restoration in 1660 have all disappeared, apart from a few which came into the hands of the first Lord Dartmouth, Master-General of the Ordnance under James

27. Viktor Hantzsch, "Beiträge zur älteren Geschichte der kurfürstlichen Kunstkammer in Dresden," *Neues Archiv für sächsische Geschichte* 23 (1902): 220–96. A history and catalogue of both collections appears in Hantzsch's *Die Landkartenbestände der Königlichen öffentlichen Bibliothek zu Dresden*, Beihefte zum Zentralblatt für Bibliothekswesen, 28 (Leipzig: O. Harrassowitz, 1904), pp. 3–27.

II. Most of them presumably perished in the fires which consumed the Palace of Whitehall and the king's private library in the 1690s. Yet the two map collections formed after 1760 by King George III—Topographical and Maritime—and preserved in the British Museum since 1828, contain numerous maps and atlases which had belonged to the Tudor and Stuart sovereigns, as well as the results of George III's own industrious collecting. These older cartographic materials were either abstracted from the Royal Library by George (then Prince of Wales) before 1757 or (more probably) had not been administered as part of it. George III regularized the distinction by his formal constitution of the separate map collections.[28]

From the late seventeenth and early eighteenth centuries we can trace a steadily increasing momentum in the recognition of maps as distinctive objects for collecting in their own right. There were various strands in this process, most of them leading back to developments in the making and use of maps.

The "geopolitical" motive, asserting itself with ever growing force, accelerated not only map production but also the collecting and preservation of maps as records of social, economic, political, or military information for use by governments. Official map reposi-

28. George F. Warner and Julius P. Gilson, *Catalogue of Western Manuscripts in the Old Royal and King's Collections*, 4 vols. (London: Trustees of the British Museum, 1921), is a list of the manuscripts transferred in 1757. R. A. Skelton has detailed the history of some of the British Museum's collections in "King George III's Maritime Collection," *British Museum Quarterly* 18 (1953): 63; "The Hydrographic Collections of the British Museum," *Journal of the Institute of Navigation* 9 (1956): 323–34, and "The Royal Map Collections of England," *Imago Mundi* 13 (1956): 181–83.

tories, growing by a process of natural accumulation, preserved the work of cadastral surveyors, of hydrographers and of military engineers. So too did privileged commercial corporations, the Hudson Bay Company, the Dutch and English East India Companies. In a period when control of public documents was lax, many "official" maps came into the personal collections of military commanders, politicians, or princes. Some such groups have thus entered the public market (for example, the Clinton papers, most of which are in the Clements Library, or the collections of Prince Eugene, now in the National Library at Vienna).[29] While it is in the personal collections of sovereigns that nearly all the national libraries of Europe have their roots, no materials in private ownership could be considered secure until they had become a public trust. As Richard Gough remarked in 1780, "A public library is the safest port"; and he added that "of all public libraries the British Museum is on the most liberal plan, deficient only in the want of a sufficient fund to furnish itself with what it may not suit the wishes or the finances of many good collectors to bestow on it."[30]

The interest of the private map collector in the seven-

29. Randolph G. Adams, *British Headquarters Maps and Sketches Used by Sir Henry Clinton while in Command of the British Forces Operating in North America During the War for Independence, 1775–1782: A Descriptive List of the Original Manuscripts and Printed Documents Now Preserved in the William L. Clements Library at the University of Michigan* (Ann Arbor: William L. Clements Library, 1928). On Prince Eugene's collections, see Wolfram Suchier, *Prinz Eugen von Savoyen als Bibliophile* (Weimar: Hempe, 1928).

30. Richard Gough, *British Topography, or an Historical Account of What Has Been Done for Illustrating the Topographical Antiquities of Great Britain and Ireland,* 2 vols. (London: Payne, T., and Son, 1780), I: xlvii.

teenth and eighteenth centuries was mainly directed to printed maps and atlases and supplied by a vigorous and expanding trade. There were still no antiquarian book-sellers, but the market in older maps and atlases was fed by libraries which came into the auction rooms. The library of Adriaan Pauw, sold at The Hague in 1654, had nine editions of Ptolemy, beginning with the Ulm edition of 1482, and most of the important atlases of the sixteenth and early seventeenth centuries.[31] At a more modest level, we find among the books of a Scottish gentleman, James Balfour of Kinnaird, which were sold at Edinburgh in 1699, seventeen atlases ranging in date from 1584 (Mercator's Ptolemy) to 1675 (Ogilby's road book), and including an Ortelius (1592), a Münster, Dutch sea atlases, and no fewer than seven atlases by Blaeu.[32]

The general trend of buying and selling, however, was toward new or recently published cartographic materials, predominantly atlases. The atlas provided the nucleus of many eighteenth-century map collections. As Professor Koeman has shown, the formation of an "atlas factice" is the hallmark of a deliberate and purposeful collector of maps.[33] The famous "Eugene Atlas," in forty-six volumes, now in the National Library of Vienna, was constructed by the Amsterdam merchant Laurens van der Hem, between 1662 and his

31. For a description of Pauw's library, see Cornelis Koeman, *Collections of Maps and Atlases in the Netherlands, Their History and Present State*, Imago Mundi Supplements, vol. 3 (Leiden: E. J. Brill, 1961), pp. 32–35.

32. *Catalogus selectissimorum in quavis lingua & facultate librorum, quorum maxima pars pertinebat ad clarissimos fratres D. D. Balfourios, Jacobum de Kinnaird Equitem et Andream, Med. Doc. Eq. Auratum* (Edinburgh: Successores Andreae Anderson, 1699).

33. Koeman, *Collections*, pp. 35–39, 64–75.

death in 1678, round Blaeu's eleven-volume atlas by "grangerizing" on a magnificent scale. It was purchased at auction by Prince Eugene in 1730.[34] The "Atlas Royal" of Augustus Strong, Elector of Saxony, put together in 1706–10, extended to nineteen volumes containing 1,400 maps.[35] Another Amsterdam merchant, Christoffel Beudecker (d. 1756), constructed a twenty-seven-volume atlas of the Seventeen Provinces in the same way.[36]

For collectors of different degrees, tastes, and purses, the map trade provided works made to order or "made to measure." This is no new practice. The manuscript atlas commissioned from a Renaissance artist of the quattrocento had its counterpart in the "superatlases" of Dutch printed wall maps, such as the "Klencke Atlas" presented to King Charles II, and that of Prince Johann Moritz.[37] The collections of engraved maps compiled "in Ptolemy's order" for customers in Vene-

34. On the "Eugene" atlas, see F. C. Wieder, ed., *Monumenta cartographica*, 5 vols. (The Hague, Martinus Nijhoff, 1925–33), 5:145–59, and H. de la Fontaine Verwey, "De atlas van Mr. Laurens van der Hem," *Maandblad Amstelodamum* 38 (1951): 85–89.

35. The "Atlas Royal" is described in Hantzsch, *Landkartenbestände*, pp. 62–64.

36. The later history of the Beudecker atlas, now in the British Museum, is given in Koeman, *Collections*, pp. 82–83.

37. The Mauritius Atlas (which was given to Frederick William, Elector of Brandenburg, about 1664) is now preserved in the German State Library in Berlin and has recently been reproduced in a facsimile edition at about one-half the size of the original. The atlas is accompanied by Egon Klemp's *Commentary on the Atlas of the Great Elector* (Stuttgart: Belser Verlag, 1971), which includes a table comparing the contents of the Mauritius and Klencke atlases as well as a third "superatlas" in the Rostock University Library.

tian and Roman shops of the sixteenth century are recalled by the atlases of various proportions and various prices offered by eighteenth-century mapsellers in the Netherlands and Germany. The early years of the century produced a crop of handbooks for map collectors, mainly of German authorship. Although mines of information for the modern bibliographer on map production of the seventeenth and eighteenth centuries, they were designed as guides for the contemporary collector. One such manual, after giving a geographically classified list of available sheet maps, set out specifications for making up twenty-four atlases varying in size from 18 sheets to 1,500 sheets (in five volumes), and in price from 3 German taler to 300.[38]

These authors did not neglect the history of cartography or early maps as an object for collecting. "Old geographers (wrote J. G. Gregorius in 1713), even if their maps are imperfect, faulty and imprecise, should not on that account be ridiculed but rather commended."[39] During the eighteenth century, however, it was less as specimens of the art of cartography than as historical documents that early maps were esteemed. The scholars most actively and perceptively interested in them were antiquaries and historians, who used them to reconstruct the geography of the past. Such pioneer

38. Johann Hübner, *Museum geographicum, das ist: ein Verzeichnis der besten Land-Charten so in Deutschland, Frankreich, England und Holland von den besten Künstlern sind gestochen worden; nebst einem Vorschlage wie daraus allerhand grosse und kleine Atlantes können gemacht werden* (Hamburg: Conrad König, 1726), pp. 315–400.

39. Johann Gottfried Gregorii, *Curieuse Gedancken van den vornemsten und accuratesten alten und neuen Land-Charten* (Frankfort and Leipzig, 1713).

work as that of Richard Gough, in analysis and collecting of early maps, made only a limited contribution to the history of cartography as an art and a science.

It was nevertheless a historian (in the broader sense) who created the first important American map collection. This was the German scholar Christoph Daniel Ebeling (1741–1817) of Hamburg. Both his major works remained uncompleted: the *Erdbeschreibung und Geschichte von Amerika* (7 vols., 1793–1816) and the *Atlas von Nordamerika* (ten sheets out of eighteen projected). The maps which he collected to support and illustrate his historical studies numbered 20,000 by 1799. In 1818, after his death, the collection was purchased by Edward Everett on Colonel Israel Thorndike's account and presented to Harvard College.[40]

Since the first half of the nineteenth century, the progress of map collecting has unfolded on parallel lines to that of book collecting. Both have exhibited the same trends—in the relations between supply and demand, in direction of movement, in the motives and methods of collectors, in the expansion of the market and its service by dealers. At the same time, the creation of map collections and the systematic cultivation of the history of cartography have marched forward side by side with increasing acceleration. Map col-

40. A brief history of the Ebeling collection and its acquisition by Harvard may be found in Mary M. Bryan, "The Harvard College Library Map Collection," Special Libraries Association, *Geography and Map Division Bulletin* 36 (1959): 4–12. The Ebeling collection, with a few exceptions, is catalogued in *A Catalogue of the Maps and Charts in the Library of Harvard University in Cambridge, Massachusetts* (Cambridge: E. W. Metcalf, 1831).

lecting in the modern period is partly a precipitation of the study of early maps considered as specimens of evolving cartographic theory and practice.

In the earlier phase, up to about 1850, Paris was a significant center, in which the seeds of future growth were sown. In the Bibliothèque Nationale, where the former royal library had been enriched by the holdings of expropriated religious houses, a separately administered Département des Cartes et Plans was created by Edme François Jomard, with a vigorous and instructed policy of acquisition.[41] At Paris was formed (among others) the important private map collection of the Baron de Walckenaer.[42] And the first steps in the methodical study of medieval and Renaissance cartography had been taken by two scholars in exile, who were also collectors: the Portuguese Viscount de Santarém, at Paris, and the Pole Joachim Lelewel, at Brussels.[43]

These activities foreshadow, in embryo, the pattern of more sophisticated and intensive effort in the retrieval, recording, and preservation of early carto-

41. Edme François Jomard, *Considérations sur l'objet et les avantages d'une collection spéciale consacrée aux cartes géographiques et aux divers branches de la géographie* (Paris: E. Duverger, 1831).

42. Charles Athenase Walckenaer, *Catalogue des livres et cartes géographiques de la bibliothèque de feu M. le baron Walckenaer* (Paris: L. Potier, 1853).

43. For a bio-bibliographical account of Santarém see Armando Cortesão, *History of Portuguese Cartography* (Lisbon: Junta de Investigações do Ultramar, 1969–), 1: 7–26. On Lelewel, see the entry under his name in the *Allgemeine Enzyklopädie der Wissenschaften und Künste* (Leipzig: Brockhaus, 1818–89), sec. 2, vol. 43, pp. 61–64, and Stanisław Warnke, *Joachim Lelewela zasługi na polu geografii* (Poznan, 1878).

graphic materials during the nineteenth and early twentieth centuries. In the first place, national map collections, emerging (in European countries) from the chrysalis of the former royal collections, were segregated, independently organized, and actively expanded. So too were institutional collections—those of the older universities, of government departments, of societies of geography. They provided homes for many important map collections formed by private individuals: for instance, those of Richard Gough (in the Bodleian) and the cartographer d'Anville (in the Bibliothèque Nationale).[44] There was much activity in the publication of inventories and catalogues. Many scholar-curators, after the model of Jomard, enriched the literature of cartographic history either by descriptive works on the map collections in their charge or by monographic studies. I may instance R. H. Major in England, Viktor Hantzsch in Saxony, Gabriel Marcel and Charles de la Roncière in France, Justin Winsor and P. Lee Phillips in the United States, F. C. Wieder in the Netherlands.

Secondly, we see the emergence of the scholar-collector whose interest was canalized into early cartography: Bodel Nijenhuis and Abraham van Stolk in the Netherlands, General von Hauslab in Austria, A. E. Nordenskiöld in Sweden (the supreme prototype of

44. Bodleian Library, *A Catalogue of the Books Relating to British Topography, and Saxon and Northern Literature, Bequeathed to the Bodleian Library in . . . 1799, by Richard Gough* (Oxford: Clarendon Press, 1814). The history and a summary catalogue of the d'Anville collection are treated in Charles du Bus, "Les collections d'Anville à la Bibliothèque Nationale," *Bulletin de la Section de Géographie* 39 (1924): 93–145.

such collectors), and—in this century—Leo Bagrow, Roberto Almagià, and George H. Beans.[45] Whether, in such men, the collector or the student takes precedence is a question not much easier to answer than whether the chicken came before the egg. Their collecting, directed by knowledge and judgment, provided the research materials for their published work on cartographic history, and the published work throws light on the collections.

The third type of nineteenth-century map collector is the omnivore. That is not really a type at all, for there was only one Sir Thomas Phillipps and there will never be another. Inevitably, many important maps were among the manuscripts swept together by the "great

45. The Bodel Nijenhuis collection, now in the University Library at Leiden, and the van Stolk collection, maintained today as a foundation, are described in Koeman, pp. 242–46 and 251–53. The Bagrow and von Hauslab collections are now in the Houghton Library at Harvard University.

The Nordenskiöld collection is now in the Helsinki University Library and a catalogue of the collection is in preparation. A description of the collection, in Nordenskiöld's own words, is printed in George Kish, "Adolf Erik Nordenskiöld (1832–1901), Historian of Science and Bibliophile," *Biblis* (1968): 171–83. An account of Nordenskiöld's cartographic collecting is given in Eino Nivanka, "Amsterdamilainen Fr. Mullerin antikvariaatti Nordenskiöldin kirjakokoelman päähankkijana" [The Booksellers Fr. Muller & Co. of Amsterdam as Chief Purveyors of Books for the Nordenskiöld Collection], *Miscellanea Bibliographica* 5 (1947): 122–33.

The John Carter Brown Library acquired those maps from the Beans collection that fell within their collecting policy. See "The George H. Beans Gift of Maps and Geographical Treatises," Brown University, John Carter Brown Library, *Report*, 1957, pp. 14–33, and subsequent reports describing additions: 1958, pp. 46–59; 1959, pp. 33–39; 1960, pp. 31–33; and 1960–65, p. 45. The remainder of the collection was sold to antiquarian book dealers and dispersed.

vellomaniac" into the Middle Hill library.[46] As, by slow degrees, they have come on the market and found homes in active libraries, these maps have joined the corpus of materials available to the map historian.

Much more significant in the story of map collecting is the appearance of a fourth type, the selective collector whose library illustrates the historical development of a particular country or region. Maps are primary documents for the maritime and commercial history and overseas expansion of European countries, for the discovery and settlement of America, for the advancing frontier and contacts between peoples. In collections constructed around these themes and dominated by the geographical setting, maps have formed an increasingly large constituent part, organically related to other materials. The great collectors of Americana—John Carter Brown, James Lenox, Edward E. Ayer, Henry R. Wagner, Thomas W. Streeter—did not (we may be sure) set out to create map collections in any strict sense. Yet it is evident that, as their collecting proceeded, most of these men found themselves acquiring maps and atlases at—or even beyond—the very limits of their regional interest. The reasons for this? There are several, to which I shall return in my next essay. The consequence? There now exists, in major libraries of the United States, a great and growing body of original materials illustrating the whole history of cartography as an art and a science, even if there are few map collections narrowly conceived.

46. A. N. L. Munby, *Portrait of an Obsession, The Life of Sir Thomas Phillipps, the World's Greatest Book Collector*, adapted by Nicolas Barker from the five volumes of Phillipps Studies (New York: Putnam's, 1967).

The significance and scope of these materials are strikingly exemplified in the impressive map exhibition organized at Baltimore in 1952 under the title *The World Encompassed;* and the catalogue of that exhibition, as an indispensable reference book, has had hard wear in map rooms all over the world.[47] For another illustration of the point, let us look at De Ricci's *Census of Medieval and Renaissance Manuscripts* and Philip Hamer's *Guide to Archives and Manuscripts* (i.e. to collections) in the U.S.A.[48] Under the rubric "Maps," De Ricci's indexes have many references; in Hamer's index, by contrast, you will find "Circuses" but not "Cartography," "Maltsters" but not "Maps."

The Ayer Collection in the Newberry Library demonstrates the range and variety—both geographical and chronological—of the map materials which a collection of Americana may legitimately admit. It is fair to say that they illustrate the evolution of cosmographic ideas and of cartographic practice from classical antiquity to the American Revolution.[49]

Into the same almost involuntary specialization in

47. Elizabeth Baer, Lloyd A. Brown, and Dorothy E. Miner, *The World Encompassed; An Exhibition of the History of Maps Held at the Baltimore Museum of Art, October 7 to November 23, 1952* (Baltimore: Trustees of the Walters Art Gallery, 1952).

48. Seymour de Ricci and W. J. Wilson, *Census of Medieval and Renaissance Manuscripts in the United States and Canada,* 3 vols. (New York: H. W. Wilson, 1935–40), with a supplement by C. U. Faye and W. H. Bond (New York: Bibliographical Society of America, 1962), and Philip M. Hamer, ed., *A Guide to Archives and Manuscripts in the United States* (New Haven: Yale University Press, 1961).

49. Clara A. Smith, comp., *List of Manuscript Maps in the Edward E. Ayer Collection* (Chicago: The Newberry Library, 1927).

cartography were drawn also the dealers and agents who—by a process of "book brokerage"—built up the great nineteenth-century collections. Henry Stevens of Vermont had a keen eye for early maps, even if he just failed to secure the La Cosa world map and the Drake-Hondius map for America.[50] Both he and his son Henry Newton Stevens made constructive contributions to the history of cartography. Henry Harrisse, whose *Bibliotheca Americana Vetustissima* (1866) provided—and still provides—the starting point for a collector of Americana, knew as much as any man about Renaissance maps, even if (in J. A. Williamson's words) "as his knowledge increased his judgment deteriorated."[51] The catalogues of cartography issued by Stevens in London, Maisonneuve in Paris, Asher in Berlin, Hiersemann in Leipzig, Rosenthal in Munich, and (above all) Frederik Muller, collector, publisher, and bibliographer of early maps and atlases, has been described by my friend Professor Koeman.[52] Muller's work, supplemented and continued by his assistants or partners P. A. Tiele, F. C. Wieder, and A. W. M. Mensing, has (as Koeman shows) laid the foundations for the history of Dutch cartography.

These dealers, and their successors, have been the architects of many important modern map collections:

50. This episode is described in Henry Stevens, *Recollections of James Lenox and the Formation of His Library*, revised and elucidated by Victor Hugo Paltsits (New York: New York Public Library, 1951), pp. 50–51.

51. James A. Williamson, *The Voyages of John and Sebastian Cabot*, Historical Association Pamphlet no. 106 (London: Pub. for the Historical Association by G. Bell & Sons, 1937), p. 7.

52. Cornelis Koeman, "Frederik Muller: His Importance for the Study of the History of Cartography," in *Collections*, pp. 91–111.

for instance, in Europe (outside the national libraries) those of the maritime museums at Rotterdam, Amsterdam, and Greenwich; in America those of the great libraries created by the enthusiasm of private collectors and now, for the most part, by a happy dispensation conveyed to institutional ownership.

If we compare the map collector's opportunities one hundred years ago, even twenty years ago, and today, the contrast is too painful to bear explicit rehearsal. The supply to the market is progressively shrinking. The gradient of price has climbed, and is still climbing, very steeply—even more so than in the book field. Since the rise in auction values is not completely justified by relative rarity, it appears that competition between collectors is more intense (unless we are to suppose that prices are being artificially forced upward). We need not go so far back as Henry Stevens, who in 1853, at the Walckenaer sale, bid 1,000 francs (the equivalent of $200) for the La Cosa map—and lost it to the Queen of Spain for $8.00 more; what would this map fetch today? But look at the auction records for the Ulm Ptolemy of 1482—the most desirable of map incunabula, but not a really scarce book:

1884..........$85
1901..........$350
1933..........$3,000
1950..........$5,000
1965..........$28,000

And for the important Strasbourg edition of 1513:

1884..........$110
1950..........$1,000
1965..........$27,200

The outlook for the map collector today is, in my judgment, much more hopeful than these depressing statistics suggest. There are several factors in his favor.

1. Leaving aside the richest "nuggets" (to use Henry Steven's word), I doubt whether the market is yet seriously starved. We cannot expect it to be often replenished on a grand scale by the dispersal of such exceptional collections as those of Mr. Beans, Mr. Kenney, and Mr. Streeter. But it is obvious that the attics of Europe are by no means cleaned out yet. Even if some dealers have little to offer but atlas maps, the catalogues of others contain many desirable map items for the discriminating collector at acceptable prices. I have the impression that, with enhanced values for complete atlases, the practice of breaking them up—and so destroying bibliographical evidence—has passed its peak.

2. The map collector is served by a number of well-informed dealers, on whom he must largely depend for his supplies. The better ones are industrious in searching out sources, and their standard of cataloguing is often much higher than that of the auction houses, showing that they do their homework.

3. The wily collector knows, however, that the most favorable time for buying a map or atlas is *before* it is catalogued by a dealer—in fact, as near as possible to the moment when it enters the market, usually at auction. Auction catalogues are rewarding study. Most dealers' prices are fair and realistic; but if a map has passed from dealer to dealer, the ultimate price paid by the collector may include the successive markups of several dealers, not merely one. For the same reasons, the collector who has time is well advised to rummage for himself in backstreet bookstores or print shops or junk shops, rather than leave this chore to the dealer.

4. Some considerable areas of map collecting have been little worked, either in extension or in depth. There are several reasons for this: the predominantly regional interest of most map collectors, their preoccupation with the content rather than the construction and form of maps, a kind of snobbery which exalts the mapping of purely geographical discovery at the expense of other cartographic records, a feeling that "special-purpose" or thematic mapping is not the concern of a humanist. In my later essays I shall suggest a broader approach to maps as objects for study and collecting.

The neglect of certain fields by collectors has extended what I call the "interval of vulnerability." This may now be said to reach back to the beginning of the nineteenth century and even beyond. Here is a great and creative tract of cartographic history—one in which maps and atlases, original in form and content, await the collector even of moderate means, provided that the pulper has not got them first. He may form a collection to illustrate developing forms of cartographic expression (for instance, methods of representing relief, the use of tone and color); or to display the enrichment of the content of maps from the earth sciences (geology, oceanography, meteorology, and so on) or from human geography (for example, population, language, communications, occupations, disease). Some of these motives or themes will lead the collector back as far as the seventeenth century, others to the eighteenth or early nineteenth. In all of them, he will find satisfaction to the eye and the mind, provided he agrees that (let us say) the earlier mapping of geology or population is, in human history, comparable in significance to the Renaissance mapping of America.

3
The Historical Study of Early Maps: Past

AT THE OUTSET I SHOULD LIKE TO MAKE A DISTINCTION in terminology. Just as "historical geography" is not the same thing as "the history of geography," so we must not confuse "historical cartography" with "the history of cartography" (or "cartographic history"). Historical geography and cartography are procedures by which geographical circumstances of the past are reconstructed, or detected in those of the present. The history of geography and the history of cartography trace the development of man's knowledge and ideas about the earth and the graphic forms in which he has expressed them.[1]

By a very rough classification, four approaches or attitudes to an early map, considered in relation to its period, can be distinguished. A map may be regarded: (1) as material for reconstructing the physical landscape, (2) as evidence of human life and organization, (3) as an illustration of the state of geographical knowledge and thought, or (4) as a product of cartographic skill and practice. It is possible, broadly speak-

1. For other definitions of these terms, see Armando Cortesão, *History of Portuguese Cartography* (Lisbon: Junta de Investigações do Ultramar, 1969), vol. 1, p. 4.

ing, to consider the first two aspects as the province of the historical geographer (as well as the archaeologist and the social, economic or political historian), the last two as matter for the historian of science and technology. In practice, of course, it is quite arbitrary to segregate these interests. Each interlocks with the others; each has a part in the counterpoint of history. Human life and the conditions in which it develops are inseparable as subjects for study. If the historical geographer or the historian must take into account the content of a map, that is, the geographical data presented in it, he cannot extract its essential kernel of fact without regard for the limitations to which the mind, eye, and hand of the contemporary mapmaker were subject, because they determine the outward and visible forms of expression employed. They in turn are partly conditioned by social and economic factors. So the study of content and that of form mutually control and support one another.

At the same time, one or other of the four themes or attitudes tends to be dominant at different periods or in the work of different scholars, and each makes its entry at separate points in the development of cartographic studies. The relatively sophisticated critical analysis of the form or "look" of maps comes late in the story.

The historical geographer, reconstructing the physical or cultural landscape of a particular period or identifying the chronological strata which underlie that of the present, is lucky if he finds early maps among his documentary sources. If he does, he must learn how to use his tools. Historical geography goes back to Plato and Herodotus; historical cartography to Crates of Mallos, who, in the second century B.C., drew a globe to illustrate the Homeric story of the wanderings of Odysseus written down some six hundred years earlier. The earliest surviving historical maps are of the later Middle Ages. Matthew Paris, in the thirteenth century, drew a map of the Roman roads in Britain, and another of the kingdoms of the Anglo-Saxon heptarchy.[2] To a large extent, medieval *mappaemundi* are historical maps. They illustrate sacred or biblical history not only in their iconography but also in the symbolism of their formal design; they depict the legendary exploits of Alexander the Great and the provincial divisions of the Roman Empire. Some of this matter was quarried from much older maps—and it is in this light that the Roman world maps of the later imperial age, which were still preserved in monastic libraries, must have been studied by medieval cartographers.

But in the Middle Ages we also find an incipient interest in the evolution of geographical ideas as expressed in maps. Lambert of St. Omer, in the twelfth century, copied into his *Liber floridus* world maps of different types and periods (Roman and post-Roman), ascribing them to their authors.[3] Many manuscripts of Goro Dati's metrical cosmography *La Sfera* (early fif-

2. British Museum, *Four Maps of Great Britain Designed by Matthew Paris about A.D. 1250* (London: Trustees of the British Museum, 1928).

3. Maps of Lambert of St. Omer are listed in Marcel Destombes, ed., *Mappemondes, A.D. 1200–1500*, Monumenta cartographica vetustioris aevi, vol. 1, Imago Mundi Supplements, vol. 4 (Amsterdam: N. Israel, 1964), pp. 111–16. See also Richard Uhden, "Die Weltkarte des Martianus Capella," *Petermanns geographische Mitteilungen* 76 (1930): 126.

teenth century) contain a *mappamundi* of archaic type alongside a modern one.[4] Andrea Bianco's manuscript atlas of 1436 includes, on successive pages, a traditional circular *mappamundi* and a Ptolemaic world map.[5] The juxtaposition of old and new models by these authors surely indicates an intention to illustrate changing ideas about world geography, not necessarily evolving knowledge—for Bianco's map is not so very much richer in content than Ptolemy's—but the contrast between distinct concepts of the pattern of land and water and also between different ways of laying them down on a map. Here we may discern a rudimentary historical sense applied to comparative cartography. By the middle of the sixteenth century the subject was being expounded at Oxford by Richard Hakluyt, who was, he claimed, "in my publike lectures . . . the first, that produced and shewed both the olde imperfectly composed, and the new lately reformed Mappes, Globes, Spheares, and other instruments of this Art."[6]

The attitude of Renaissance geographers to Ptolemy tells a similar tale. During the fifteenth century the maps accompanying the *Geographia* were turned into Latin, copied and recopied, and (from 1477) printed in the first place to illustrate the manual of mapmaking

4. Destombes, *Mappemondes*, pp. 249–51.

5. Bianco's maps are reproduced in *Der Atlas des Andrea Bianco vom Jahre 1436, in zehn Tafeln (photographische Facsimile in der Grösse des Originals)* ed. Max Münster with fwd. by Oscar Peschel (Venice: H. F. & M. Münster, 1869); also Italian editions of 1871 and 1879.

6. "Epistle dedicatorie," in *The Principall Navigations, Voiages & Discoveries of the English Nation*, 2 vols. (1589; reprint ed., Cambridge: University Press, 1965), 1: fol. *2.

in the text and in the second place as the only available complete and consistent world atlas. The great discoveries gave a fresh impetus, from 1490 onward, to the printing of the *Geographia*—but they also progressively discredited Ptolemy's world picture and diminished its authority.

The attempts made by editors (from 1482) to bring Ptolemy's atlas up to date by adding modern maps or (as in the 1511 edition) by correcting his maps from recent information were rightly chastised as ramshackle and confusing, by Martin Waldseemüller.[7] His Strasbourg edition of 1513 is the first to separate the modern maps (constructed by regional surveys) from the ancient ones, and so to ascribe a historical character to those of Ptolemy, which reflected the geographical knowledge and the cartographic practice of the second century. Mercator's edition of 1578 reverted to the original practice (for the first time since the Rome edition exactly one hundred years earlier)* of printing the Ptolemaic maps alone, without any modern supplement. This underlined their purely historical interest as a facsimile of a classical atlas.

It was the sudden explosion of knowledge and the abruptly accelerated transformation of men's ideas about the world that fostered the historical sense of Renaissance geographers. This was expressed in the contradistinction of "ancient" and "modern" geography which was to dictate the pattern of atlas publica-

* Except, of course, the 1490 Rome edition, which contained maps printed from the same plates.—Ed.

7. Claudius Ptolemaeus, *Geographia* (Strasbourg, 1513), fol. L2v.

tion from Ortelius to the nineteenth century. But between these poles there was little or no attempt to illustrate the evolution of geographical thought and of mapmaking in the flux of time.

The historical geography which flourished in the sixteenth and seventeenth centuries was also confined, in the main, to the illustration of biblical and classical history. These were the themes of the maps in the historical atlas—the *Parergon*—which Ortelius added to his *Theatrum* from 1579. This too was the motive for the facsimile publication of the Peutinger Table planned by Konrad Peutinger and accomplished by Marcus Welser and Ortelius. The Elizabethan antiquaries who drew maps with Anglo-Saxon place names, or to illustrate the political divisions of Anglo-Saxon England, found no early maps as compilation materials.

Medieval maps were neither unknown nor completely overlooked. The "Gough map," a road map of England made in the middle of the fourteenth century, still served as a source for English and German cartographers in the middle of the sixteenth.[8] The map of Central Europe known by the name of Cardinal Nicholas of Cusa, probably drawn before the cardinal's death in 1464 and first engraved not many years later, was reprinted from the original fifteenth-century plate at Basel in 1531.[9] In each of these cases, however, the

8. E. J. S. Parsons, *The Map of Great Britain circa A.D. 1360 Known as the Gough Map* (Oxford: University Press, 1958), p. 15.

9. For a description of this map, with an extensive bibliography, see Marcel Destombes, *Catalogue des cartes gravées au XVe siècle* (Paris: Union Géographique Internationale, 1952), pp. 79–81.

rate of change in the topographical situation represented was too small to discourage the sixteenth-century cartographer from reproducing it in a contemporary map. Again, in the catalogue of cartographers which Ortelius prefixed to his atlas—a priceless source for the modern map historian—almost all the names are those of men whose work Ortelius had incorporated in his own maps. In other words, he set out to produce something more like *Who's Who in America* than the *Dictionary of American Biography;* and we are not justified in supposing his motive to have been historical.

Before the eighteenth century, in fact, medieval maps are seldom cited or referred to, and very few were reproduced as historical documents. There are occasional curious exceptions. Ramusio refers to maps "two or three hundred years old" illustrating Marco Polo's travels. The earliest historical maps of America are those drawn about 1600 in Iceland and Denmark in an attempt to locate on the modern map the Norse discoveries and settlements of the eleventh century recorded in sagas and annals. The most elaborate of these maps (by Bishop H. P. Resen, 1605) twice quotes among its sources (for example, for "the old sailing route to Iceland") an "ancient Icelandic map, rudely drawn some centuries ago (ante aliquot antennis)." This "ancient map" has—I need hardly say—disappeared.[10]

Another old and tattered map was found among

10. The sixteenth- and seventeenth-century maps incorporating information from ancient Icelandic records are summarized in R. A. Skelton, Thomas E. Marston, and George D. Painter, *The Vinland Map and the Tartar Relation* (New Haven and London: Yale University Press, 1965): 199–208. Resen's map is reproduced in plate xix, opposite p. 147.

family papers in his house at Venice by Antonio Zeno, who in 1558 had it engraved, with additions, to illustrate his account of voyages into the Atlantic said to have been made by his forbears shortly before 1400. The literary evidence for these voyages is equivocal; but the original MS map at least—though now lost—was genuine. It must have been a map of the North drawn in the later fifteenth century by Donnus Nicolaus Germanus or Henricus Martellus Germanus. Antonio Zeno's engraved version was, however, designed not as a facsimile reproduction but as an illustration of history, which he felt free to adapt to his text.[11]

A more conventional type of reproduction is represented by the publication in 1611 of Marino Sanudo's tract *Liber Secretorum Fidelium Crucis*, completed in 1321 for the promotion of a crusade. The manuscripts contain a series of maps and charts by Sanudo or Petrus Vesconte, and the seventeenth-century editor Jacques Bongars had them engraved with the printed text.[12]

Thus, during the sixteenth and seventeenth centuries the neglect of medieval cartography was not total;

11. The Zeno map is described in R. A. Skelton et al., *Vinland Map*, pp. 197–99, and is reproduced on p. 193.

12. Marino Sanudo, *Liber secretorum fidelium crucis* . . . , ed. Jacques Bongars (Hanover: Typis Wechelianis, apud heredes Ioannis Aubrii, 1611).

Three of these maps (the Holy Land, Eastern Mediterranean area, and Jerusalem) are reproduced in an English translation of part of Sanudo's *Secretorum: Part XIV of Book III of Marino Sanuto's Secrets for True Crusaders to Help Them to Recover the Holy Land* (London: Palestine Pilgrims' Text Society, 1896). See also Konrad Kretschmer, "Marino Sanudo der Ältere und die Karten des Petrus Vesconte," Gesellschaft für Erdkunde zu Berlin, *Zeitschrift* 26 (1891): 352–70.

but the exceptions prove the rule. A map of the late Middle Ages might be used in compiling a current map if it contained a residuum of still valid geographical fact; or it might be drawn upon as a source for historical information. Significantly, it was never contemplated or analyzed as an artefact, and no notice was taken of the techniques by which it had been constructed and drawn. The study of cartographic expression and form as a mode of communication had not yet begun. The consequent slighting of medieval mapmaking, though understandable, exemplifies the general want of interest in cartographic history, as a continuous process, before the eighteenth century.

In 1856 Johann Georg Kohl, lecturing at the Smithsonian Institution, made some severe comments on this neglect. "The history of geographical maps [he wrote] . . . remained a perfect blank until our days."[13] Thirteen years later he returned to the charge: "Even those maps and charts, which had been spared by all-destroying time, were scarcely noticed by the historians and geographers of the last century, sharing the neglect with which . . . Gothic and other medieval monuments were regarded."[14]

These animadversions hardly do justice to the handful of eighteenth-century geographers and antiquaries

13. Johann Georg Kohl, "Substance of a Lecture Delivered at the Smithsonian Institution on a Collection of the Charts and Maps of America," in Smithsonian Institution, *Annual Report of the Board of Regents . . . 1856* (Washington: Cornelius Wendell, 1857), p. 95.

14. *History of the Discovery of Maine*, Maine Historical Society, Documentary History of the State of Maine, vol. 1 (Portland: Bailey and Noyes, 1869), p. 25.

who turned their eyes on early maps. Some of the manuals for the map collector even attempted a systematic, if naïve, summary of cartographic history; Johann Gottfried Gregorius, in 1713, offered his readers a chronological catalogue of geographers and mapmakers beginning with Moses (whom he called "the first geographer") and ending with Homann.[15] There were critical studies of medieval Arabic maps and globes by orientalists. Venetian scholars, aware that (as G. F. Zanetti wrote in 1758) "to a nation trading by land and sea, like ours, cosmographic knowledge is more necessary than to any other," located and described the nautical charts drawn at Venice in the fourteenth and fifteenth centuries.[16] Research into the origins of imaginary Atlantic islands led the French geographer Philippe Buache to examine fifteenth- and sixteenth-century maps showing Antillia and Friesland.[17]

There were engraved reproductions of early maps: the Peutinger Table again (Nuremberg, 1753), Bianco's world map (Venice, 1783), Behaim's globe (Nuremberg, 1730 and 1778), medieval maps of Britain (by Richard Gough in 1780).[18] There were also restrikes

15. Gregorii, *Curieuse Gedancken.*

16. Girolamo Francesco Zanetti, *Dell' origine di alcune arti principali appresso i Viniziani, libri due* (Venice, 1758).

17. Philippe Buache, "Dissertation sur l'île Antillia," *Mémoires sur l'Amérique et sur l'Afrique donnés au mois d'avril 1752* (n.p., 1752).

18. *Peutingeriana tabula itineraria quae in Augusta Bibliotheca Vindobonensi nunc servatur numini majestatique Mariae Theresiae regniae . . . dicata a F. C. de Scheyb* (Vienna, 1753). Vincenzo Antonio Formaleoni, *Saggio sulla nautica antica de' Veneziani, con una illustrazione d'alcune carte idrografiche antiche* [*by Bianco*] *della biblioteca di S. Marco, che dimostrano l'Isole-Antille prima della scoperta di Cristo-*

from old plates or blocks. An early fifteenth-century world map on a niello plate was acquired by Cardinal Stefano Borgia in 1794, and impressions were made from casts and published in 1797; pulls from the woodcut Turkish map of Hadji Ahmed engraved at Venice in 1560 were taken for the first time in 1795.[19]

Some systematic works produced in the eighteenth century are still used with profit by the modern student. For instance, from Nuremberg, Johann Gabriel Doppelmayr's historical account of the mathematicians and instrument makers of his native city (1730), and Georg Martin Raidel's bibliographical commentary on the manuscripts and printed editions of Ptolemy's *Geographia* (1737).[20] Richard Gough's *British Topogra-*

---

foro Colombo (Venice: presso l'autore, 1783). The Behaim globe was reproduced in Johann Gabriel Doppelmayr, *Historische Nachricht von den nürnbergischen Mathematicis und Künstlern, welche fast von dreyen seculis her, durch ihre Schrifften und Kunst-Bemühungen die Mathematic und mehreste Künste in Nürnberg* (Nuremberg: P. C. Monath, 1730); and Christoph Gottlieb Murr, *Diplomatische Geschichte des portugesischen berühmten Ritters Martin Behaims* (Nuremberg: J. E. Zeh, 1778). Richard Gough, *British Topography, or an Historical Account of What Has Been Done for Illustrating the Topographical Antiquities of Great Britain and Ireland*, 2 vols. (London: Payne, T., and Son, 1780).

19. Description, history, list of reproductions, and bibliography of the Borgia map appear in Destombes, *Mappemondes*, pp. 239–41. A history and reproduction of the Hadji Ahmed map appear in George Kish, *The Suppressed Turkish Map of 1560* (Ann Arbor: William L. Clements Library, 1957).

20. Doppelmayr, *Historische Nachricht;* Georg Martin Raidel, *Commentatio critico-literaria de Claudii Ptolemaei Geographia, eiusque codicibus tam manu-scriptis quam typis expressis* (Nuremberg: typis et sumtibus haeredum Felseckerianorum, 1737).

phy (1780) contains the first comparative study of medieval maps of Great Britain.[21]

Gough's approach was that of the antiquary. The eager interest in the remains of classical antiquity, characteristic of this period, explains the success with which the most extraordinary of map forgeries was launched. A map of Roman Britain supposed to have been drawn, from an ancient original, by the fourteenth-century monk Richard of Cirencester, was published in 1747 by the archaeologist William Stukeley from a drawing sent to him from Copenhagen by a young Dane, Charles Bertram. It was accepted into the canon of historical sources for the geography of Roman Britain for over a century before Bertram's fraud was exposed. Until the 1920s, the Ordnance Survey still marked Roman stations laid down from Richard of Cirencester's map and confirmed by no other source.[22]

In spite of these scattered manifestations of curiosity about early maps during the eighteenth century, Kohl had some reason to contrast its relatively tempered enthusiasm with the frenzied activity of his own day,

21. Gough, *British Topography*, vol. 1, pp. 57–86.

22. William Stukeley, *An Account of Richard of Cirencester, Monk of Westminster, and of His Works; with His Antient Map of Roman Brittain and of the Itinerary Thereof. Read at the Antiquarian Society, March 18, 1756* (London: Richard Hett, 1757). Bertram's forgery, under the title "Ricardi monachi westmonasteriensis commentarioli geographici de situ Brittaniae" appears in Stukeley's *Itinerarium curiosum: Or, an Account of the Antiquities, and Remarkable Curiosities in Nature or Art, Observed in Travels Through Great Britain*, 2d ed., 2 vols. (London: Baker & Leigh, 1776), 2:79–108. For an account of Bertram and the exposure of his fraud, see his entry in the *Dictionary of National Biography*, vol. 4, pp. 412–13.

when (as he wrote in 1856) it was "quite a common thing to edit old maps and globes" on both sides of the Atlantic.[23]

This release of energy had been effected by pressures of various kinds. Geography was one of the disciplines which, in the early nineteenth century, hived off from polymathic or encyclopedic science, creating their own institutions; and the newly formed societies of geography were hungry for work. A stricter approach to records of the past was served by an immense effort in cataloguing and listing, in editing, printing, and reproduction. In the age of the Gothic revival (as Kohl observed), medieval cartography—no less than medieval architecture—became a respectable subject for preservation and study. It was seen to furnish the historical background for the great discoveries in the East and the West. Interest in the discovery and exploration of America received an impulse from Alexander von Humboldt and C. C. Rafn, and this gathered momentum throughout the century.[24] Among the new nations of the Americas there was a lively historical consciousness, exemplified in the United States by a mushroom growth of historical societies. Political disputes about sovereignty or frontiers were frequent, and the testimony of early maps was invoked.

23. Kohl, "Lecture," p. 97.

24. Alexander von Humboldt, *Examen critique de l'histoire de la géographie du nouveau continent, et des progrès de l'astronomie nautique aux 15me et 16me siècles* (Paris: Gide, 1836–39), and Carl Christian Rafn, *Antiquitates americanae; sive, Scriptores septentrionales rerum ante-columbianarum in America. Samling af de i nordens oldskrifter indeholdte efterretninger om de gamle Nordboers opdagelsesriser til America fra det 10de til det 14de aarhundrede* (Copenhagen: typis officinae Schultzianae, 1837).

History, especially in an age which identified it with progress, could be seen as in continuous movement, accelerating as it approached the present. The historian admitted a concern not only with the remoter past, but equally with the immediate past, divided from the future only by a point in time. To illustrate this historical attitude in relation to the collection and study of maps of the past, I cannot resist another quotation from Kohl's lecture of 1856, though in this case the lesson he taught has been all but forgotten:

> With Columbus commenced the *hydrographical* discovery and chartography of America. The *geological* discovery and chartography of America began only a few years ago. . . . If we should collect and preserve the one class, there is no reason why we should not likewise provide an asylum for the other. . . . [This] holds good also with regard to the botanical, zoological, magnetical, ethnographical and other numerous classes of maps. Each of them . . . has inaugurated a discovery of America in a new sense.[25]

In the intellectual climate created by these conditions there appeared a number of original and productive scholars. Like other figures from the age of romanticism, they now seem rather larger than life. They produced monumental publications under extreme difficulties; they conducted gigantic wars of words; and they laid the foundations for the history of cartography.

The keynote was struck by a more sedate character, Placido Zurla, a Camaldolese monk (of Fra Mauro's

25. Kohl, "Lecture," p. 124.

monastery at Murano) who became a cardinal. Early maps—he wrote in 1818—are "monuments of high interest which reveal at a glance the state of geographical knowledge and the art of representation characteristic of different nations and cultures." This attitude to maps as geographical artefacts found relaxed and objective expression in Zurla's well-documented studies of Venetian cartography in the Middle Ages and Renaissance—studies which still hold water.[26]

But it was at Paris that the real ferment in cartographic history set in, after the foundation of the Société de Géographie in 1821 and the creation of a map division in the Bibliothèque Royale in 1828 under Edme François Jomard. Jomard was a versatile scholar who, as a young *ingénieur-géographe*, had served in the scientific staff of Napoleon's expedition to Egypt in 1798. In the Bibliothèque Royale, working in difficult circumstances, he instituted a vigorous policy of acquiring medieval maps, in original or in facsimile, with the intention (as expressed in his words) first of "provoking a search for early maps still unknown and bringing them out of the dust and oblivion in which they are buried," and second "of facilitating comparative study of the ancient manuscript maps scattered throughout Europe."

In 1834, Manuel Francisco de Barros y Souza, 2d Viscount of Santarém, left Portugal and its "fatal vor-

26. Placido Zurla, "Appendice sulle antiche mappe idrogeografiche lavorate in Venezia," in his *Di Marco Polo e degli altri viaggiatori veneziani piu illustri dissertazioni*, 2 vols. (Venice: Presso G. G. Fuchs, 1818), 2:299. Other studies by Zurla include *Il mappamondo di Fra Mauro Camaldolese, descritto ed illustra . . .* (Venice, 1806), and *Dissertazione intorno ai viaggi e scoperte settentrionali di Nicolò ed Antonio fratelli Zeni* (Venice: Dalle stampe Zerletti, 1808).

tex of revolutions and political reaction" to settle in Paris, where he lived until his death in 1856. His interest in early maps seems to have been first aroused when, in an interval of his political career, he was director of the Portuguese archives from 1824 to 1827; and he pursued it with passion in exile. To this dedicated man we owe (with much else) the first known use—and perhaps the coining—of the word "cartography" (in a private letter of 1839).[27]

Santarém's most important published work had its roots in nationalism and represents an early—though not the first—invocation of old maps in a political dispute.[28] In 1840 the sovereignty over Casamance in Senegal was at issue between Portugal and France. Santarém's memoir in support of the Portuguese case set out to demonstrate "the incontestable priority of the discovery of Africa beyond Cape Bojador by the Portuguese." An atlas of map facsimiles, in twenty-one sheets, accompanied the Portuguese edition of the memoir in 1841; for the French edition of 1842 the atlas was enlarged to thirty-one sheets. After this, Santarém's studies in early cartographic history and his reproduction project outran his original motive. In its definitive form, in 1849, the *Atlas* contained sev-

27. Quoted in Cortesão, *History*, vol. 1, pp. 4–5.

28. Some early instances of the use of maps in boundary disputes are mentioned in P. Lee Phillips, "The Value of Maps in Boundary Disputes, Especially in Connection with Venezuela and British Guiana," American Historical Association, *Annual Report for 1896*, vol. 1, pp. 457–62, and Charles Cheney Hyde, "Maps as Evidence in International Boundary Disputes," *American Journal of International Law* 27 (1933): 311–16. The Ebeling maps (see above, p. 52) were used, beginning in 1829, for the same purpose. See Kimball C. Elkins, "Harvard Library and the Northeastern Boundary Dispute," *Harvard Library Bulletin* 6 (1952): 255–63.

enty-seven sheets reproducing 180 subjects and embracing the whole range of cosmography and cartography from the eighth century to the beginning of the seventeenth.[29]

Santarém's reproductions are hand-drawn copies printed by lithography, many in color. Although not (in the strict sense) facsimiles, they are still indispensable as the only generally available corpus of medieval *mappaemundi* drawn from a great number of European libraries.

When Santarém presented his first *Atlas* to the Société de Géographie in March 1842, Jomard laid claim to priority, in intemperate terms, on the ground that "for several years" he had been preparing a collection of medieval maps in facsimile and that he would be exposed to a charge of plagiarism by Santarém's prior publication. Both men took time off to argue this ridiculous issue with heat to prolixity.[30]

Jomard's own atlas, *Les monuments de la géographie* (Paris: Duprat), appeared serially from 1842 to 1862. It contained no more than thirty maps or globes on eighty-one plates; but the subjects were selected with excellent judgment (in Jomard's view, "surviving maps are not a true index to contemporary knowledge"), and the standard of reproduction, especially the color work, is higher than Santarém's. Jomard's expressed purpose was to record originals "exposed to the ravages of time," since "to publish is to save them from

29. For biographical data and a survey of the various editions of Santarém's Atlas, see Cortesão, *History*, vol. 1, pp. 7–26.

30. Details of the disagreement are given by Cortesão, ibid., pp. 29–32.

early ruin." It is in fact evident that in his time more names could be read on the La Cosa map (which he reproduced) than can be deciphered today in ultraviolet or infrared light. There is perhaps a lesson for us here.

Both Santarém and Jomard bore the production costs of these atlases from their own purses. Joachim Lelewel, historian and head of the Polish government in exile at Brussels, had more slender financial resources, and he drew and engraved with his own hand the plates for his facsimile atlas published in 1852.[31] Of the three scholars, it was Lelewel, working with demonic energy and near-fanaticism, who provided his atlas with the most penetrating and complete accompanying text, to which the map reproductions are subsidiary. Lelewel's work cannot be safely ignored even today, in spite of occasional naïveté arising from his difficulties in obtaining research materials, particularly copies of tracings of maps. Lelewel contrasted his "precarious position and resources" with those of Jomard, whom he reproached for slowness in publication. Nevertheless, in his *chambrette* or "little room" (as he called it) at Brussels, Lelewel accumulated a remarkable working library, including 316 atlas volumes, with editions of Ptolemy, seven of Ortelius, and four Blaeu atlases ranging in size from four to eleven volumes.[32] I suspect the "chambrette" to have been a romantic hyper-

31. Joachim Lelewel, *Géographie du moyen âge. . . . Accompagné d'atlas et de cartes dans chaque volume*, 5 vols. (Brussels: Ve et J. Pilliet, 1852–57).

32. For a description of Lelewel's collection, see Mikołaj Dzikowski, *Zbior kartograficzny Uniwersyteckiej Biblioteke Publiczny w Wilnie* (Wilno: Zakłady graficzne "Znicz," 1932).

bole, like Lelewel's practice of wearing workman's clothing, or George Orwell's habit of drinking tea out of the saucer.

Each of these three scholars, in his own way, tackled the problems defined by Jomard—that of "bringing early maps out of oblivion" and that of "facilitating comparative study of them." To this pioneer phase belongs also the work of J. G. Kohl, whom I have already quoted. A native of Bremen, he was an active, observant, and articulate traveler and prolific author. The significance of his contribution to the historical study of cartography in this country can hardly be exaggerated.

To serve his research into the history of geography, Kohl assembled in Dresden hand-drawn copies of early maps illustrating American discovery and exploration. In 1854 he came to the United States with them, and two years later he obtained a government appropriation to prepare a series of these copies "as the foundation of an elaborate catalogue of the early maps of the American continent." This was the year of his Smithsonian lecture, in which he laid down the principles for organizing a map collection. Working for the most part in Harvard College library, Kohl also prepared for the U.S. Coast Survey memoirs on the early cartography of the American coasts. By 1858 none of these materials had been published; "The Government was almost bankrupt . . . and Dr. Kohl went home to Germany almost broken-hearted."[33] Eleven years later, at the instance of R. H. Major of the British Museum, he undertook for the Maine Historical Society his *His-*

33. Charles Deane, [Remarks on the death of Dr. J. G. Kohl] *Massachusetts Historical Society Proceedings* 16 (1879): 382.

tory of the Discovery of Maine, a classic in its perceptive use of early maps. The map copies, nearly 500 in number, remained in the Department of State until 1886, when they were catalogued at Harvard under Justin Winsor's direction; and in 1903 they were moved to the Library of Congress.[34]

A parallel exercise in copying, under modern conditions, was carried out three-quarters of a century later by Louis C. Karpinski, professor of mathematics at the University of Michigan. Using a sabbatical leave in 1926–27, he had photographs made of some 800 manuscript maps of America in libraries and archives of France, Spain, and Portugal. Positive sets of the "Karpinski Series of Reproductions" were subscribed for by six American libraries, and Karpinski's own reference materials are preserved in the Yale University Map Room.[35]

In 1856 Kohl had emphasized the interest and vulnerability of maps from the immediate past: "the first surveys of counties . . . organized within the memory of people still living are, in some cases, no longer extant." Karpinski put this principle into practice. In his valuable *Bibliography of the Printed Maps of Michigan* (Lansing: Michigan Historical Commission, 1931), he listed "county, township, village, and city maps to 1880."

34. Justin Winsor, *The Kohl Collection (Now in the Library of Congress) of Maps Relating to America* (Washington: Government Printing Office, 1904).

35. The project was outlined in Louis C. Karpinski, "Manuscript Maps Relating to American History in French, Spanish, and Portuguese Archives," *American Historical Review* 33 (1927–28): 328–30. Complete sets of the reproductions were deposited with the Library of Congress, the William L. Clements Library, the Huntington Library, the New York Public Library, and the Newberry Library.

I have dwelt on the labors of the pioneers because they formulated and faced the central problems in the comparative study and use of early maps as historical documents—problems with which we still live. They asked the right preliminary questions: "Where are the maps?" and "How can they be examined by the student in visual juxtaposition?" Their work illustrates the daunting difficulties which the scholar had to overcome before the day of photography and while there were still few published library catalogues of books and manuscripts, even fewer of maps.

Jomard and Kohl perceived the need for concentration of map resources, with central repositories strong in originals and reproductions. This principle is still valid. We must suppose that in the eyes of Santarém and Jomard the publication of facsimiles had a higher priority and more urgency than that of inventories. In taking this view, these men were doubtless influenced by the apprehension, expressed by Jomard, that the originals might in time suffer loss or deterioration. In the circumstances of today we cannot criticize this argument, although the recent tendency has been to put the compiling of inventories of resources first. In any case, Santarém and Jomard pointed the way to further development in the recording and reproduction of early maps. The facsimile atlas had come to stay; and, as a systematic anthology of characteristic specimens, it has become a standard tool for the cartographic historian.[36]

The majority of facsimile atlases before 1900 repro-

36. A number of recent facsimiles are reviewed in Ena L. Yonge, "Facsimile Atlases and Related Material: A Summary Survey," *Geographical Review* 53 (1963): 440–46. The field of facsimile publishing is surveyed by Cornelis Koeman, who

duced specimens from a single collection or a regional group of collections, selected either as historical documents, particularly for the discovery of America, or as samples of cartography. Thus Friedrich Kunstmann's *Atlas zur Entdeckungsgeschichte Amerikas* (Munich, 1859) presented manuscript maps in the Bavarian libraries, with lithographic plates from hand-drawn copies. Libraries of Venice, Florence, and Milan supplied the originals for the set of photographs of fifteen manuscript atlases and maps of the fourteenth and fifteenth centuries published by F. Ongania of Venice (1875–81).[37] Gabriel Marcel edited three photolithographic collections reproducing medieval and Renaissance maps in the Bibliothèque Nationale (1883, 1886, 1896), and another on the discovery and exploration of America, drawing on Paris libraries (1893).[38] The Amsterdam

also supplies a classification for facsimiles and discusses reproduction techniques: "An Increase in Facsimile Reprints," *Imago Mundi* 18 (1964): 87–88. Some of the same ground, but with more emphasis on older facsimiles, is covered by Walter W. Ristow, "Recent Facsimile Maps and Atlases," *Quarterly Journal of the Library of Congress* 24 (1967): 213–29, which was reprinted under the title "New Maps From Old: Trends in Cartographic Facsimile Publishing" in *Canadian Cartographer* 5 (1968): 1–17. A current bibliography of facsimile maps is Walter W. Ristow and Mary E. Graziani, *Facsimiles of Rare Historical Maps: A List of Reproductions for Sale by Various Publishers and Distributors*, 3d ed., with supplements (Washington: Government Printing Office, 1968).

37. Ferdinando Ongania, *Raccolta di mappamondi e carte nautiche del XII al XVI secolo* (Venice, 1875–81).

38. Gabriel Marcel, *Choix de documents géographiques conservés à la Bibliothèque Nationale* (Paris, 1883); *Recueil de portulans* (Paris, 1886); *Choix de cartes et de mappamondes des XIVe et XVe siècles* (Paris: E. Leroux, 1896); and *Reproductions de cartes & de globes relatifs à la découverte de l'Amerique du XVIe au XVIIIe siècle* (Paris: E. Leroux, 1893).

bookseller Frederick Muller put out a series under the title *Remarkable Maps of the XV–XVII Centuries* (Amsterdam, 1894–97), from originals in the Leiden university library.

Few compilers of facsimile atlases before 1900 cast their net so widely as did Santarém. The exceptions are the facsimile atlases edited by Konrad Kretschmer and Adolf Erik Nordenskiöld. In Kretschmer's *Die Entdeckung Amerikas* (Berlin, 1892), all maps are redrawn for reproduction. This provides an effective historical synthesis, giving the atlas continuing value as a graphic index of early cartography and as a teaching aid.

Nordenskiöld was distinguished as a geologist and as the explorer who first traversed the Northeast Passage. He also did more than any other scholar to break down the difficulties created for the student of early cartography by the dispersal and often the inaccessibility of his materials. His *Facsimile Atlas to the Early History of Cartography* (Stockholm, 1889) reproduced "the most important maps printed in the XV and XVI centuries." His *Periplus* (Stockholm, 1897), though it declares itself "an essay on the early history of charts and sailing directions," also illustrates the evolution of mapping of the world and its regions, reproducing manuscript as well as printed maps. If we take account of the extraordinary range of reference, the critical judgment and systematic arrangement of the text, the correctness of historical proportion—all based on solid geographic scholarship—and the high quality of the photolithography, it is safe to say that these two atlases will never be superseded on our reference shelves, though they may be subject to correction in detail.

Three more important facsimile collections with regional reference, published early in this century, have to be mentioned: E. L. Stevenson's facsimiles of twelve Renaissance maps "illustrating early history and discovery in America" (1903–6), A. B. Hulbert's *Crown Collection* of reproductions of manuscript maps in the Public Record Office and British Museum (1904–9), and Count P. G. Teleki's atlas of early maps of Japan (1909).[39]

It is an unfortunate fact that most of the facsimile atlases to which we must still refer are already themselves *rariora*. Some—like Stevenson and Hulbert—were produced in very small editions for restricted distribution. All are scarce and highly priced in the market. Paradoxical as it may seem, photo-offset reprinting of the facsimile atlases has now begun.[40]

The output of facsimiles has not slackened since the middle of the late nineteenth century; today in fact the tide is flowing more strongly than ever before. Comparative study of early maps is made easier, even if the need for an index of map facsimiles is increasingly felt. The publication of inventories and catalogues of map collections, surveys of map holdings, and

39. Edward Luther Stevenson, *Maps Illustrating Early Discovery and Exploration in America, 1502–1530* (New Brunswick, N. J., 1903–6); Archer Butler Hulbert, *The Crown Collection of Photographs of American Maps* (Cleveland: Arthur H. Clark Co., 1904–9); and Count Pál Teleki, *Atlas zur Geschichte der Kartographie der japanischen Inseln* (Budapest: K. W. Hiersemann, 1909).

40. For example, Nordenskiöld's *Facsimile-atlas* (New York: Kraus Reprint Corp., 1961) and *Periplus* (New York: Burt Franklin, 1967), and Emerson D. Fite and Archibald Freeman's *A Book of Old Maps Delineating American History* (New York: Dover Publications, 1969).

union lists has not proceeded so briskly. This is understandable. A catalogue demands a more sustained effort than the selection of originals for reproduction: it is a serious business, not to be undertaken lightly. The question "Where are the maps?" is still a troublesome one; so is the question "Where are the other materials for studying them?"

Suppose yourself a student of the history of cartography fifty years ago, say in the year 1916, surveying the published aids at your disposal without leaving your home town. You would notice, to begin with, that the map collections of national libraries had done very little to tell you what they contained. Only one—the British Museum—had published complete catalogues of its holdings (manuscript maps 1844–61, printed maps 1885). From the Library of Congress, under the direction of Philip Lee Phillips, had come lists of its maps of America and the incomparable *List of Geographical Atlases* (volumes 1–4). National archives had done better, with map catalogues from the Rijksarchief (The Hague) and the Archivo General de Indias (Seville). You would find invaluable surveys, with locations, of early manuscript maps preserved in Italy (Uzielli and Amat, 1882) and in Germany (W. Ruge, 1904–16); and one of globes in Italy (Fiorini, 1898).[41] Catalogues or guides were available for maps

41. British Museum, *Catalogue of the Manuscript Maps, Charts, and Plans, and of the Topographical Drawings in the British Museum*, 3 vols. (London: Printed by order of the Trustees, 1844–61) and *Catalogue of the Printed Maps, Plans, and Charts in the British Museum*, 2 vols. (London: Printed by order of the Trustees, 1885). Philip Lee Phillips, *A List of Maps of America in the Library of Congress* (Washington: Government Printing Office, 1901) and *A List of Geographi-*

in various libraries or archives of Spain, Italy, Germany, the Netherlands, Sweden, and the United States.

The student of 1916 would find himself much better served by critical catalogues or lists of surviving maps or charts distinguished by type and form. Konrad Miller's *Mappaemundi* (Stuttgart, 1895–98) reduced all the known medieval world maps to order by synoptic description and genetic classification; this was a momentous achievement which has stood the test of time. Later printed world maps were listed by Nordenskiöld (the *Facsimile Atlas*). Nautical charts and sailing directions of the Middle Ages and Renaissance had been covered by the catalogues of Nordenskiöld (the *Periplus*), Walter Behrmann (1906) and Konrad Kretsch-

cal Atlases in the Library of Congress (Washington: Government Printing Office, 1909–20). A continuation of this list, edited by Clara Egli LeGear is in progress, and vols. 5 and 6 have been published. The Dutch archives catalog is *Inventaris der verzameling kaarten berustende in het Rijks-archief* (The Hague: M. Nijhoff, 1867–71). A series of catalogues compiled by Pedro Torres Lanzas lists the cartographic holding of the Archivo General de Indias in Seville. Catalogues were published for Virreinato de Buenos-Aires (1900), Filipinas (1897), Guatemala, San Salvador, Honduras, Nicaragua y Costa-Rica (1903), Panamá, Santa Fé y Quito (1904), México y Floridas (1900), and Perú y Chile (1906). Gustavo Uzielli and P. Amat di S. Filippo, *Mappamondi, carte nautiche, portolani ed altri monumenti cartografici specialmente italiani dei secoli XIII-XVII* (Rome: Società [geografica italiana], 1882; reprinted Amsterdam: Meridian Publishing Co., 1967); Walter Ruge, "Aelteres kartographisches Material in deutschen Bibliotheken," K. Gesellschaft der Wissenschaften zu Göttingen, Philologisch-historische klasse, *Nachrichten*, 1904, pp. 1–69; 1906, pp. 1–39; 1911, pp. 35–166; 1916, pp. 1–128; and Matteo Fiorini, *Sfere terrestri e celesti di autore italiano, oppure fatte o conservate in Italia* (Rome: La Società geografica italiana, 1899).

mer (1909).[42] The great atlas output of the Netherlands was recorded in the bibliography by P. A. Tiele (1884);[43] many documentary sources for their map industry in the sixteenth and seventeenth centuries had been printed, notably the letters of Ortelius, the Plantin archives, and the records of the Amsterdam book trade. Numerous mapmakers (German, Italian, French, Flemish, and Dutch) were the subject of bio-bibliographies. There was an extensive literature surveying the cartography of particular countries or regions. Specially significant, for America, are the great bibliographical and analytical works of Henry Harrisse, from the first in 1866 (*Biblioteca Americana Vetustissima*) to the last in 1900 (*Découverte et évolution cartographique de Terre-Neuve*), and in I. N. Phelps Stokes's *Iconography of Manhattan Island* (New York, 1916), with F. C. Wieder as cartographic editor. All these aids to locating early cartographic materials are still in use today unrevised, although collections have expanded and many important maps have migrated. Surprisingly, the British Museum is still the only national library that has yet published its full map catalogue. In 1967, the museum, using a photo-abstracting process, brought out a new edition of its catalogue of printed maps, recording its holdings at the end of 1964; the two vol-

42. Walter Behrmann, *Über die niederdeutschen Seebücher des fünfzehnten und sechzehnten Jahrhunderts* (Hamburg: L. Friederischen, 1906), and Konrad Kretschmer, *Die italienische Portolane des Mittelalters: ein Beitrag zur Geschichte der Kartographie und Nautik* (Berlin, 1909; reprinted Hildesheim: G. Olms, 1962).

43. Pieter Anton Tiele, *Nederlandsche bibliographie van land- en volkenkunde* (F. Muller, 1884; reprinted Amsterdam: Theatrum Orbis Terrarum, 1966).

umes of the original 1885 edition are now swollen to fifteen.[44]

From all or most of these systematic works of the nineteenth and early twentieth centuries we learn what maps of a particular class exist, and where they are to be found. I should say, perhaps, "what early maps were known to these authors," who—for want of regular published catalogues of maps—had to search them out from library catalogues, by correspondence, and by visits to innumerable repositories. We are in slightly—but not much—better condition today.

The works which I have just enumerated stand on the reference shelves of map rooms today as they did half a century ago. Imperfect though they often seem in the light of modern knowledge, they have been neither revised nor superseded. Many have become so scarce and high-priced that enterprising publishers in the Netherlands, Germany, and other countries have recently reprinted them by photolithography, without the addition or correction which they need. If the student is not to be misled, he must still supplement them by reference to newer literature, scattered through many publications and often traced only by the eye of experience. Anyone who enjoys the fun of the hunt will agree that research should not be made too easy; but this is not to excuse avoidable expense of time and labor. If continued demand for an old reference book justifies a reprint, it also justifies some effort in bringing it up to date. The recent reprint of Ganong's classic

44. British Museum, *Catalogue of Printed Maps, Charts, and Plans*, photolithographic ed., 15 vols. (London: Trustees of the British Museum, 1967).

papers on early Canadian cartography shows how this can and should be done.[45] That this is not common practice throws some light on the present state of studies in cartographic history and of their publication.

The three-quarters of a century from Santarém's first publication to World War I saw the production of a great literature of monographs on early maps and on aspects of their history. I can only refer briefly to a few of the themes which it illuminated and on which it said the first (and occasionally the last) word.

1. The patterns and purposes of the medieval world map, and its Roman origins.
2. Changing ideas on the shape and size of the earth, as they affected the mathematical construction of maps. The history of map projections.
3. The growth of topographical mapping, from instrumental techniques of observation and survey.
4. The enrichment of geographical knowledge by discovery and exploration; its reflection in cartography.
5. The development of methods of navigation, and its effect on the design of sea charts.
6. Evolution of cartographic representation and design: symbols, expression of relief, and so on.
7. The application of printing techniques to cartography.
8. The development of the map trade.

45. William Francis Ganong, *Crucial Maps in the Early Cartography and Place-Nomenclature of the Atlantic Coast of Canada*, with an introduction, commentary, and map notes by Theodore E. Layng (Toronto: University of Toronto Press, 1964).

Generally, the interest of students in this period stopped short of the nineteenth century. With very few exceptions, it did not extend to the initiation of scientific national survey, to the beginnings of thematic mapping, or to the profusion of what I may call "social" mapping in Europe and America during the nineteenth century. The admonition of Kohl, which I quoted earlier, was ignored.

With that reservation, the great and varied body of intensive study which reached publication during the period remains of lasting value. Whether the approach was technical, historical, bibliographical, or iconographic, it firmly established the habit of looking at an early map as an artefact for the visual communication "of significant spatial ideas, facts and interrelationships."[46] Supported by the new resources of photography, it placed the comparative study of cartographic history on a solid foundation.

46. Arthur H. Robinson, "The Potential Contribution of Cartography in Liberal Education," *Cartographer* 2 (1965): 1. A more complete version of this paper, including a list of the slides that were used to illustrate it, appears in John F. Lounsbury, ed., *Geography in Undergraduate Liberal Education: A Report of the Geography in Liberal Education Project* (Washington: Association of American Geographers, 1965), pp. 34–47.

4

The Historical Study of Early Maps: Present and Future

IN RESEARCH AND PUBLICATION ON EARLY MAPS AND cartographic history, the period between the two world wars was one of solid accomplishment and marked advance. It was also, in retrospect, a period of frustration, in that—despite various promising beginnings—it failed to provide the subject with a firm general base, secure lines of communication, and an accepted methodology. The growing recognition that the history of cartography was an interdisciplinary study implied that it was also extradisciplinary. What is everybody's business is nobody's business; and this was a baby which no established institution has proved willing to adopt.

These are the seeds of the present situation in our field of study, to which prewar developments are a necessary prelude.

In his large treatise *Die Kartenwissenschaft* (1921–25) the German geographer Max Eckert systematically analyzed the character and evolution of different types of map and established genetic principles for their formal study. The great variety and wide range of the examples which he quoted in illustration of his argument make the book a mine of reference for the cartographic historian.[1] But it is not a synthesis, and Eckert himself held that a general history of cartography was not yet possible, partly because he could not see any one scholar with the necessary competence and scope, partly because the materials for it were still being assembled and were (as yet) far from complete. He emphasized the urgent need for a coordination of work by individuals and for bibliographical organization of the materials. He himself had suffered from difficulties of communication. His book was planned and written during and immediately after a world war; his references are mainly to maps available in German libraries, and the "genetic facsimile atlas" which he projected had to be abandoned because of the difficulty of obtaining the map reproductions needed from libraries in Paris and London.

Eckert's judgment on the state of the subject was to be justified by its further progress in the 1920s and 1930s. In these years the literature was enriched by copious publication, in a variety of forms, resulting from the work of many scholars in Europe, Asia, and America. The landmarks—both before and since World War II—are still the facsimile atlases, for which superior techniques of reproduction by collotype or offset lithography using fine screens had become available. Their planning, and the selection of examples, followed one or other of three principles: (1) to illustrate the mapping of a region or country (the largest class),

1. Max Eckert, *Die Kartenwissenschaft: Forschungen und Grundlagen zu einer Kartographie als Wissenschaft*, 2 vols. (Berlin and Leipzig: Vereinigung Wissenschaftlicher Verleger, 1921–25).

(2) to illustrate the map production of a country, or (3) to reproduce maps in a particular collection or collections.

In the first class appeared the most lavish of all facsimile atlases, the *Monumenta cartographica Africae et Aegypti* edited by F. C. Wieder for Prince Youssouf Kamal of Egypt, who had it printed in sixteen great volumes or "fascicules" (Cairo, 1926–51), each weighing about thirty pounds.[2] Other regionally organized facsimile atlases of this period were those of Italy (Almagià 1929), the Sudetenland (Brandt 1930–33), Bohemia (Švambera and Šalamon 1930–37), the Ukraine (Kordt 1931), Denmark, Iceland, and the Faeroes (Nörlund 1942–44), and the British Isles (Crone 1961). For America, there was the well-known and still useful *A Book of Old Maps* by Emerson Fite and Archibald Freeman (1926); maps in Spanish archives were reproduced in atlases edited by Admiral J. F. Guillén (1942) and the Duke of Alba and others in 1951; and the American West has been treated in the five volumes of Carl I. Wheat (1957–63).[3]

2. The individual maps in these elephant folio volumes are described by Clara E. LeGear in *A List of Geographical Atlases in the Library of Congress* (Washington: Library of Congress, 1963), 6:426–58.

3. Roberto Almagià, *Monumenta Italiae cartographica* (Florence: Istituto geografico militare, 1929); Bernhard Brandt, *Kartographische Denkmäler der Sudetenländer*, 10 vols. (Prague: Kommissionsverlag K. André, 1930–33); Václav Švambera and Bedřich Šalamon, *Monumenta cartographica Bohemiae*, 2 vols. (Prague, 1930–37); Veniamin A. Kordt, *Monumenta cartographica Ucrainae* (Kiev, 1931); Niels Erik Nörlund, *Danmarks kortlaegning, Islands kortlaegning, Faeroernes kortlaegning*, 3 vols., issued separately (Copenhagen: Munksgaard, 1942–44); Gerald R. Crone, *Early Maps of the British Isles A.D. 1000–A.D. 1579* (London: Royal

The map production of the Netherlands in the seventeenth century was illustrated in F. C. Wieder's *Monumenta Cartographica* (The Hague, 1925–33)—apparently the first atlas to use this now familiar form of title, and that of Portugal from the fifteenth to the eighteenth centuries by the *Portugaliae Monumenta Cartographica* of Armando Cortesão (Lisbon, 1960–62). One more splendid postwar example exemplifies the reproduction of maps from a single repository: *Monumenta Cartographica Vaticana*, by Roberto Almagià (Vatican City, 1944–55).

I have enumerated these facsimile publications for two reasons. First, their systematic plan and elaborate commentaries give them the character of monographs or histories. Second, the scope or theme of the great majority is the national state; and this is true also of many publications in monograph form.

This predominant division of the field of study along (so to speak) political boundaries is intelligible and obviously convenient for authors. It is acceptable to the students so long as it is understood to apply to the collection and classification of the *materials* only, and not to the general *history* of cartography, which must be a synthesis wide and flexible enough to embrace and to use other methods of classification as well. Even if, for instance, regional mapping, topographical or cadastral

---

Geographical Society, 1961); Julio F. Guillén y Tato, *Monumenta chartográfica Indiana* (Madrid, 1942); Duke of Alba (and others), *Mapas españoles de América, siglos XV–XVI* (Madrid: Real Academia de la Historica, 1951); and Carl I. Wheat, *Mapping the Transmississippi West, 1540–1861*, 5 vols. (San Francisco: Institute of Historical Cartography, 1957–63).

in character, may submit to political delimitation, such categories as nautical charts or military maps transcend it and demand grouping by form and content.

The facsimile atlases were costly undertakings for which the necessary financial backing could, as a rule, only be provided from governmental funds; and this goes some way toward explaining their national bias. Apart from them, the interwar period saw some notable attempts at synthesis on broader principles. The fundamental work in the classification and elucidation of Arabic cartography was done by Konrad Miller (1926–31) when he was in his eighties, and on the manuscript tradition of Ptolemy's *Geographia* by Joseph Fischer (1931). The systematization of medieval *mappaemundi*, begun by Miller, was further refined by M. C. Andrews (1926) and Richard Uhden (1931) using a wider range of examples.[4] The evolution of instrumental methods in land-surveying and topographical mapping was investigated. The history of navigation and of nautical cartography was illustrated by studies or catalogues of charts and sailing directions—although here again the grouping has tended to follow national lines (Italian, Catalan, Portuguese, Spanish, Dutch, Swedish, English).

4. Konrad Miller, *Mappae arabicae; arabische Welt- und Länderkarten des 9.–13. jahrhunderts* 6 vols. (Stuttgart: Selbstverlag des Herausgebers, 1926–31); Joseph Fischer, *C. Ptolomaei Geographiae Codex Urbinas graecus 82*, 2 vols. in 4 (Leiden: E. J. Brill; Leipzig: O. Harrassowitz, 1932); Michael Corbet Andrews, "The Study and Classification of Medieval Mappae Mundi," *Archaeologia* 75 (1925–26): 61–76; and Richard Uhden, "Zur Herkunft und Systematik der mittelalterlichen Weltkarten," *Geographische Zeitschrift* 37 (1931): 321–40.

Otherwise, in spite of the wealth of research and publication between the wars, it must be said that the comparative study of map construction and of map forms, from a historical standpoint, was somewhat neglected. Professor Eduard Imhof, writing in 1964, had some reason to complain of "the yawning gaps in work on cartographic history."

> Many publications confine themselves to descriptions of maps and to bibliographical, chronological, and biographical data. . . . We would much rather have more precise information on the topographical instruments and methods of survey, on graphic techniques, on the degree of accuracy, the novelty of content, contemporary influences, and (above all) on the author's originality.[5]

Nor did this period make much contribution to the recording of resources. No regional guides appeared, and very few union lists; I may mention the union list of atlases in libraries of Chicago, published by the University of Chicago in 1936.[6] The facsimile atlases give much information about the whereabouts of maps and atlases; but this necessary job can be adequately done by publications more modest in form and much less expensive to produce.

If the initiative for more substantial publications had been taken before World War I mainly by private in-

5. Eduard Imhof, "Beiträge zur Geschichte der topographischen Kartographie," *International Yearbook of Cartography* 4 (1964): 130.

6. University of Chicago Libraries, *Atlases in Libraries of Chicago; A Bibliography and Union Checklist* (Chicago: University of Chicago Libraries, 1936).

dividuals and after the war by bodies at the national level, some weakness in international organization and coordination of these studies may be inferred. Since the first International Geographical Congress met in 1871, geography had its organ for intercommunication. Successive congresses (at nominal four-yearly intervals) showed abundant interest in the history of cartography. The climax was reached at the Amsterdam Congress in 1938, when, of the thirty-five papers read in the section on historical geography and the history of geography, twenty-eight dealt with early maps, for the most part at a distinguished level of scholarship. The Geneva Congress of 1908 was the first one to appoint a Commission for the Reproduction of Early Maps. Reappointed at the Rome Congress in 1913, the commission then lapsed until it was revived by the Cambridge Congress in 1928, with Almagià (whose *Monumenta Italiae Cartographica* had just appeared) as chairman; and it remained in being until World War II. The commission (renamed the Commission on Early Maps) was reappointed by successive postwar congresses, from that of Lisbon in 1949, until it was eventually dissolved at the London Congress in 1964. It was succeeded by a working party which carried on the projects. In 1965 the members of the working party brought out the first volume of the projected four-volume catalogue of medieval maps; another is being prepared for the press, and the research needed to complete the materials for the other two is in progress.[7]

7. International Geographical Union, *Monumenta cartographica vetustioris aevi A.D. 1200–1500* (Amsterdam: N. Israel, 1964—).

From its beginning in 1908, the commission, despite its name, wavered between reproduction of selected early maps and inventorization of surviving specimens. In 1928 it undertook the preparation of a facsimile atlas to be entitled *Monumenta Cartographica Europae*. By 1938 it had still got no further than planning, although its discussions and the publicity given to them had provoked independent activity in various countries; and in that year's congress an alternative scheme was proposed by Leo Bagrow, who undertook to publish the resulting atlas himself.[8] Both projects were checked by the war. The postwar commission, judging that the preparation of an inventory should precede that of a corpus of facsimiles, undertook a catalogue of all maps produced before the year 1500.

The prewar commissions undoubtedly did much to stimulate studies and communication between scholars on early cartography and to promote the publication of papers; but they were sterile in the facsimile publication which they had undertaken. Bagrow, whose intervention in 1938 I have mentioned, played a much more effective part in organizing international cooperation in the study of cartographic history and in getting the results into print.

Leo Bagrow was a Russian scholar "with fire in his

8. See [Leo Bagrow], "Übersicht der Tätigkeit internationaler Geographenkongresse auf dem Gebiete der Geschichte der Kartographie," *Imago Mundi* 1 (1935): 65–68, and the reports of the 1938 Congress held in Amsterdam in *Imago Mundi* 3 (1939): 100–102.

belly" who dedicated himself to this field of study with passionate energy throughout his life, taking Nordenskiöld as his model. He served in the Hydrographic Department of the Russian Navy until 1918, when he emigrated to Berlin. For forty years this was the base for his intensive activity in searching out early maps, in collecting, in study, and in publication. In May 1945 a Swedish plane took him, his wife, and their pet sparrow to Stockholm, where—under the patronage of the king of Sweden—he lived and worked until his death in 1957 at the age of seventy-six.[9]

Bagrow's peculiar qualities of character and scholarship—enthusiasm and ability to communicate it, formidable pertinacity, a genial but masterful disposition—equipped him to project and promote work in his chosen subject. His research was thorough and penetrating; he had a ranging mind which led him, on one hand, to travel incessantly in search of old maps—at the age of seventy-five he visited Ethiopia—and, on the other, to conceive imaginative plans for the publication of collective works, facsimiles, and monographs. Bagrow's own interest lay in the early map as a visual artefact and in its formal qualities and evolution. Of the many fields in which he worked, those which he cultivated most intensively were the printed maps of the fifteenth and sixteenth centuries, Russian cartography, and the maps of non-European peoples.

It is not surprising that, in the 1930s and after World War II, Bagrow's influence was strong and pervasive.

9. For details of Bagrow's life and work, see his obituary (by R. A. Skelton) in *Imago Mundi* 14 (1959): 5–12. A bibliography listing seventy-three items is appended.

It encouraged students to concentrate their attention on the face of the map; it stimulated and evoked much research into cartographic history and its documentation; it developed a system of communication with scholars of all countries across national barriers.

The foundation of the annual *Imago Mundi* in 1935 was Bagrow's principal legacy. It was designed as "a journal which might in time become an international center of information." To an interdisciplinary study, such as cartography, a journal is particularly necessary as a current and continuous register of work done or in progress. Because the substantive data depicted in maps may be drawn from many fields, the study of maps considered in respect to their content has a centrifugal tendency. The results are often published in places where the formal aspects of a map—the expressive terms by which it makes its communication—are of little interest. Anyone concerned with the map as an artefact, that is, with its formal qualities, requires a consolidated bibliographical record of work so dispersed.

This point is illustrated by inspection of the first annual bibliography published in the first issue of *Imago Mundi*. The bibliography contains some two hundred items, of which about three-quarters are articles in periodicals. Some were published (as we might expect) in journals of geography, history, local history, geodesy and survey, hydrography and navigation, the history of science. But articles on early maps appeared in many other less obvious quarters—periodicals devoted to physical science, biology, agriculture, magnetism, economics, political science, art history, oriental studies, the classics, archaeology, printing history, bibliography and library science, archives. If classified by their subject content only, many of these materials would be overlooked by the cartographic historian. It is only a form classification which brings them to his notice.

By 1939, three issues of *Imago Mundi* had been brought out by Bagrow, and ten more followed after the war and before his death in 1957. The journal is now owned by a nonprofit company registered in London, managed by an international committee, edited at Utrecht, published at Amsterdam, and precariously maintained, during the past few years, by a subsidy from a foundation. The nineteen issues published to date from a corpus of varied information on early maps to which, in a large map collection such as that of a national library, reference has to be made almost daily.*

In the postwar scene, then, correlation of effort at the international level has continued within the same narrow limits as in the 1930s, and not less falteringly. It has not been very much more apparent at the national and regional levels. At the same time, to judge from published work, the activity of individual students has not slackened; rather, it has increased and intensified in both volume and variety. This activity rests on no agreed methodology or standards; it is poorly sustained by catalogues of resources and other aids, so that com-

* *The Map Collectors' Series*, begun in 1963 by R. V. Tooley and still comprising ten numbers a year, provides another useful source of biographical, bibliographical, and monographic information on the history of cartography. Also appearing in 1963 was the first series of the *Theatrum Orbis Terrarum* atlases in facsimile. Five series have now been published, all with valuable bibliographical introductions, many of which were written by Dr. Skelton.—Ed.

parative studies run the hazard of incompleteness, and it can be criticized by a geographer as being out of balance. The volume of this activity indicates a healthy energy; but this is not enough to raise the whole edifice of cartographic history, which should be built on intensive and extensive study of surviving examples, with analysis emerging in synthesis and leading to generalization on the evolution of the form and content of early maps.

If the research potential of students and of available study materials is to be fully realized, it is obvious that a great deal remains to be done. I propose to outline a number of tasks, which, taken together, seem to make up a reasonable program. To the question "How are they to be tackled?" I suggest some answers; but to the question "Who is to tackle them?" I offer none at this stage.

1. The first task is the establishment of the basic critical principles, methods, and practice to be observed in studying an early map. These apply both to visual analysis of the map, so that we extract all it has to yield without the need for a second inspection, and also to description of the map in a form intelligible to other students and permitting comparison with other maps described in the same form.

Visual analysis embraces all the formal characteristics of the map: its mathematical construction (projection and grid), degree of accuracy (having regard to the scale), cartographic design, representational technique and symbols, lettering and decoration, mode of preparation (drawn or engraved?), method of engraving, state of the plate or block. Since the construction and accuracy of a map are controlled by the processes of compilation or survey from which it was produced, these processes need to be identified and analyzed.

Description applies to the results of visual inspection, that is, to formal characteristics. It embraces the content of the map (for example, geographical detail or place names) only if this throws light on the history of the plate or block. Appropriate and agreed terminology is to be prescribed.

For this first task, conference between specialists, or consultation, will be necessary so that the principles adopted are in step with general usage outside cartography, for instance by students of graphic art. The end product will be a series of (perhaps three) compact manuals of guidance on such subjects as survey and compilation, drawing and design, engraving and printing. I put this task first because work is going on all the time, and the sooner we have a common language of analysis, identification, and description in which to speak to one another, the better. Moreover, the statement of the elementary conventional practices now in use—such as I have suggested—will provide a base from which new and unconventional techniques of map study and analysis can be developed, evaluated and (conceivably) incorporated in standard usage.

Two recent examples come to mind. First, Dr. Alexander B. Taylor's adaptation of the "calculus of variants" (as practiced in textual criticism) to the collation of place names, and so to the determining of genetic relationships, in a series of early maps.[10] Sec-

10. A. B. Taylor, "Name Studies in Sixteenth Century Scottish Maps," *Imago Mundi* 19 (1965): 81–99.

ond, the experimental method demonstrated by Dr. Waldo Tobler (of the University of Michigan) for testing the hypothetical construction of some medieval maps on known projections.[11] Other possibilities may suggest themselves, for instance: Has any bibliographer of printed maps yet made use of the Hinman Collating Machine?

We can hardly now call the use of the distortion grid experimental; but Professor Imhof has made it clear that the usual procedures for ascertaining the character and precision of the survey behind an early topographical map are hardly strict enough.[12]

2. The second task is the recording of map resources, with locations. This is of course a gigantic and endless labor; and we should break it down into operations which are practicable and would produce maximum benefit to students in the shortest time. From this point of view, the most useful types of record seem to be:

a. The survey-guide to a single map collection or to the map collections in a country or region.

b. The union catalogue of maps and atlases held in a country or region.

c. The union catalogue or checklist of maps or atlases of a particular type held in a country or region.

d. The catalogue of any map collection complete enough within its field to be useful for general ref-

11. Waldo R. Tobler, "Medieval Distortions: The Projections of Ancient Maps," *Annals of the Association of American Geographers* 56 (1966): 351–60.

12. Imhof, "Beiträge."

erence, for instance, that of a national library or a special collection.

e. The catalogue of a collection of manuscript maps, archival or artificial.

All such catalogues need to be kept up to date, and we should not bite off more than we can chew. Some older catalogues or union lists badly need revision.

3. The third task is the recording of the location and movement of important maps and atlases, or groups of them. When migration occurs, or has occurred, it should be published, and correction should be made to any already published catalogue.

4. The fourth task is the recording of the location of documentary source materials on the history of cartography. Surveys of manuscripts already in progress can be drawn on. The catalogues of publishers are an important source for the history of map plates and of the map trade.

5. The fifth task is the survey of the present state of studies in early cartography, or on particular phases of it, so as to draw attention to specific neglected topics or parts of the subject field.

In suggesting gaps in coverage, I could spend many words. I will, simply and subjectively, pick out a very few which seem to me interesting and important:

a. Reexamination of regional maps from the sixteenth century onward, using more exact methods of control; and the drawing of growth curves to illustrate "the rate of cartographic progress."

b. Evolution of thematic or topical elements in cartography before the nineteenth century; the emergence of map types.

c. The history of European military survey.

d. The history of national survey departments: military, topographical, hydrographic.

e. The development of map-printing techniques in the social environment.

f. The relationship, technical and commercial, between the map houses of London and Amsterdam in the seventeenth century. The structure and practices of the map trade at various periods.

g. The compiling of biographical dictionaries.

6. The sixth task is to compile and publish bibliographies of literature: general, special, and current.

7. The seventh is to prepare an index of published facsimiles.

8. The eighth is to promote the conservation of our heritage of early maps. This is of course the responsibility of libraries and other repositories, but two measures suggest themselves: an agreed code of conservation, and the establishment of an international photographic archive (as proposed by Ena L. Yonge in 1960).[13]

9. The ninth task is to promote regional cooperation between libraries. This is also a responsibility of libraries, but students have much to gain from it.

10. The tenth task is to develop the effective use of early maps and map collections by printed aids to students, by exhibitions, or by other forms of publicity.

The commonest form of request in a map room is for particular information rather than for a particular map.

13. "Preservation of Our Heritage–Imperative!" Congresso Internacional de História dos Descobrimentos, *Actas* (Lisbon: Comissão Executiva das Comemorações do V Centenário da Morte do Infante D. Henrique, 1961), vol. 2, pp. 411–21.

Few visitors to map rooms know what kind of questions a map will answer, let alone what map will answer the particular question they may be asking, or to what extent and within what limits of precision its answer may be trusted. From this and from other evidence we may infer that a very considerable portion of the map resources available in any active collection remains unexplored by persons—by no means unsophisticated persons—to whom they could be useful. In other words, these resources remain inert and await projection.

This tentative program of ten tasks does not exhaust the possibilities. I have, for instance, said nothing about possible projected facsimile atlases (which can, I think, take care of themselves), or about publication (where the scholar may and should exert some influence). All the same, the program points to ways by which the methodology of the study of an early map can be developed in the context of its period, and by which comparative study of early maps may lead us to a correct historical synthesis. There is here plenty of work for institutions and individuals; and it is work which deserves coordination and cooperation.

Each map is the product of intellectual, manual, and technical processes. In studying it, in order to extract its kernel of fact, we follow—perhaps unconsciously—its growth in the mind and under the hand of its maker and in the engraving and printing shops. In interpreting the map, the student must understand, and be able to evaluate, elements from different disciplines.

The need for this fairly sophisticated process of appreciation is not always realized by persons interested only in the content of the map. A geographer has said that "even geographers themselves, with their proprietary interest in the field of cartography, are generally far less concerned with the more complex communicative and humanistic aspects of cartography than . . . with the basic substantive material that appears on maps."[14]

The study of form, no less than that of content, in an early map transcends the bounds of disciplines. In this sense, cartographic history is an interdisciplinary study. "There are few results of man's activities that so closely parallel man's interests and intellectual capabilities as the map."[15] In the study of maps, coordination and cooperation—though not less necessary than in other fields—are much retarded. "Everybody's business is nobody's business." I wish that I could present a more positive report on the present state of these studies. But (in Jakob Burckhardt's words) "this is a story which instead of ending in periods keeps ending in question marks."

14. Arthur H. Robinson, "The Potential Contribution of Cartography in Liberal Education," *Cartographer* 2 (1965): 7.
15. Ibid., p. 5.

Raleigh Ashlin Skelton (1906-1970): A Bibliography of Published Works

COMPILED BY ROBERT W. KARROW, JR.

1946

(With G. R. Crone) "English Collections of Voyages and Travels, 1625–1846." In *Richard Hakluyt and His Successors: A Volume Issued to Commemorate the Centenary of the Hakluyt Society*, edited by Edward Lynam, pp. 63–140. Hakluyt Society Publications, 2d ser., vol. 93. London: The Hakluyt Society.

1947

Discussion of "Richard Hakluyt," by E. G. R. Taylor. *Geographical Journal* 109:173.

Review of *A Book of Voyages*, edited by Patrick O'Brian. *Geographical Journal* 110:237–38.

1948

"Maps and Map Making." In *New Universal Encyclopedia.*

1949

"Colonial Exhibition, British Museum." *Burlington Magazine* 91:253.

Discussion of "Some Medieval Theories About the Nile," by O. G. S. Crawford. *Geographical Journal* 114:25.

(With A. Codazzi) "International Geographical Congress, Lisbon, 1949." *Imago Mundi* 6:93–95.

"Ludolf's Map of Abyssinia (1683)." In *The Red Sea and*

Adjacent Countries at the Close of the Seventeenth Century, edited by Sir William Foster, pp. 182–85. Hakluyt Society Publications, 2d ser., vol. 100. London: The Hakluyt Society.

Review of *Der Behaim-Globus zu Nürnberg*, by Oswald Muris. *Geographical Journal* 113: 121–23.

Review of *Maps, Their Care, Repair, and Preservation*, by Clara E. LeGear. *Journal of Documentation* 5: 174–75.

Review of *Trade Winds: A Study of the British Overseas Trade during the French Wars, 1793–1815*, edited by C. Northcote Parkinson. *Geographical Journal* 114: 215–17.

1950

Articles on Jean Baptiste Bourguignon d'Anville, Jodocus Hondius, Gerardus Mercator, Abraham Ortelius, and others. In *Chambers's Encyclopedia*. New York: Oxford University Press. (Also in later editions.)

"Edward William O'Flaherty Lynam" (obituary). *Geographical Journal* 115: 135–36.

"Exploration." In *Oxford Junior Encyclopedia*.

"Pieter van den Keere." *The Library*, 5th ser., 5: 130–32.

Reviews of *The Fry and Jefferson Map of Virginia and Maryland: A Facsimile of the First Edition in the Tracy W. McGregor Library*, with an introduction by Dumas Malone, and *A Topographical Description of the Dominions of the United States of America*, by Thomas Pownall, edited by Lois Mulkearn. *The Library*, 5th ser., 5: 282–83.

1951

"Bishop Leslie's Maps of Scotland, 1578." *Imago Mundi* 7: 103–6.

"Decoration and Design in Maps Before 1700." *Graphis* 7: 400–413.

"Early Maps and the Printer." *Printing Review* 57:39–40.
"Jean Le Clerc's Atlas of France, 1619." *British Museum Quarterly* 16:60–61.
"John Norden's Map of Surrey." *British Museum Quarterly* 16:61–62.
"Proposals for an Historical Catalogue of Early Globes." *Archives Internationales d'Histoire des Sciences* 4:731–34.
"Watermarks and the Study of Early Maps." Review of *Watermarks: Mainly of the Seventeenth and Eighteenth Centuries*, by Edward Heawood. *Geographical Journal* 117:221–22.

1952

Decorative Printed Maps of the Fifteenth to Eighteenth Centuries: A Revised Edition of "Old Decorative Maps and Charts," by A. L. Humphreys. London, New York: Staples Press.
Review of *Captain James Cook, R.N., F.R.S.: A Bibliographical Excursion*, by Maurice Holmes. *The Library*, 5th ser., 7:220–21.
"Tudor Map Engravers." Review of *Engraving in England in the Sixteenth and Seventeenth Centuries*, by Arthur M. Hind. *Geographical Journal* 118:489–91.
"Tudor Town Plans in John Speed's *Theatre*." *Archaeological Journal* 108:109–20.

1953

"Captain Cook's Wife." *Mariner's Mirror* 39:62–63.
"Explorers' Maps. 1. The North-East Passage." *Geographical Magazine* 26:119–31.
"Explorers' Maps. 2. The North-West Passage: Frobisher to Parry." *Geographical Magazine* 26:192–205.
"King George III's Maritime Collection." *British Museum Quarterly* 18:63–64.

"Les relations anglaises de Gerard Mercator." *Bulletin de la Société Royale de Géographie d'Anvers* 66:3–10.

Review of *The Founding of the Second British Empire, 1763–1793*, vol. 1, *Discovery and Revolution*, by Vincent T. Harlow. *Geographical Journal* 119:480–81.

Reviews of *Maps*, Library Resources in the Greater London Area, 1, by Philip M. Paris; *Règles adoptées pour la conservation des collections et la rédaction des catalogues;* and *Vedettes du catalogue analytique*, published by the Département des Cartes et Plans, Bibliothèque Nationale. *Journal of Documentation* 9:226–28.

"Visscher's View of London." *British Museum Quarterly* 18:65.

1954

"Captain James Cook as a Hydrographer." *Mariner's Mirror* 40:92–119.

"The Conservation of Maps." Society of Local Archivists, *Bulletin* 14:13–19.

"Explorers' Maps. 3. Cathay or a New World? The Discovery of America from Columbus to Magellan." *Geographical Magazine* 26:519–33.

"Explorers' Maps. 4. The Portuguese Sea-way to the Indies." *Geographical Magazine* 26:610–23.

"Explorers' Maps. 5. European Rivalry for the Spice Islands." *Geographical Magazine* 26:627–38.

"Explorers' Maps. 6. Marco Polo and His Successors." *Geographical Magazine* 27:267–80.

"Explorers' Maps. 7. The Far East in the Sixteenth and Seventeenth Centuries." *Geographical Magazine* 27:339–51.

Introduction to "Liste alter Globen in Österreich–List of Early Globes in Austria," by Robert Haardt. *Der Globusfreund* 3:7–36.

Review of *Maps and Their Makers*, by G. R. Crone. *Journal of Documentation* 10:137–38.

1955

(Editor) *Charts and Views Drawn by Cook and His Officers and Reproduced from the Original Manuscripts.* Cambridge: For the Hakluyt Society at the University Press. (Portfolio to accompany *The Journals of Captain James Cook on His Voyages of Discovery*, edited by J. C. Beaglehole.)

Discussion of "Medieval Land Surveying and Topographical Maps," by Derek J. Price. *Geographical Journal* 121:7,10.

"Explorers' Maps. 8. The Spanish in the Pacific." *Geographical Magazine* 27:543–56.

"Explorers' Maps. 9. The Dutch Quest of the South-land in the Seventeenth Century." *Geographical Magazine* 28:28–39.

"Explorers' Maps. 10. James Cook and the Mapping of the Pacific." *Geographical Magazine* 28:95–106.

Review of *The Mathematical Practitioners of Tudor and Stuart England*, by E. G. R. Taylor. *Mariner's Mirror* 41:342–43.

1956

Discussion of "The Ordnance Survey and the Public," by J. C. T. Willis. *Geographical Journal* 122:155.

"Explorers' Maps. 11. The New World in the Sixteenth Century." *Geographical Magazine* 28:439–50.

"Explorers' Maps. 12. North America from Sea to Sea, 1600–1800." *Geographical Magazine* 28:489–501.

"Explorers' Maps. 13. The Rivers of Africa." *Geographical Magazine* 29:149–62.

"Explorers' Maps. 14. The Polar Regions in the Nineteenth Century." *Geographical Magazine* 29:187–200.

(With G. P. B. Naish) "Explorers' Ships—I." *Geographical Magazine* 29:374–87.

"The Hydrographic Collections of the British Museum." *Journal of the Institute of Navigation* 9:323–34.

Remarks on being awarded the Gill Memorial by the Royal Geographical Society. *Geographical Journal* 122:409.

Review of *English County Maps: The Identification, Cataloguing, and Physical Care of a Collection*, by R. J. Lee. *Journal of Documentation* 12:124–25.

"The Royal Map Collections of England." *Imago Mundi* 13:181–83.

"Stuart Map Engraving." Review of *Engraving in England in the Sixteenth and Seventeenth Centuries*, part 2, *The Reign of James I*, by Arthur M. Hind. *Geographical Journal* 122:100–101.

1957

(With F. G. Emmison) "*The Description of Essex*, by John Norden, 1594." *Geographical Journal* 123:37–41.

(With G. P. B. Naish) "Explorers' Ships—II." *Geographical Magazine* 29:436–46.

"Land Surveyors." *Journal of the Society of Archivists* 1:172.

"Mr. Leo Bagrow" (obituary). *Nature* 180:684.

"Restoration of Documents." *The Library* 5th ser. 12:58. (A letter, commenting on a review of *Manuscripts and Documents: Their Deterioration and Restoration*, by W. J. Barrow.)

"Two English Maps of the Sixteenth Century." *British Museum Quarterly* 21:1–2.

1958

"Cartography [ca. 1750–ca. 1850]." In *A History of Technology*, edited by Charles Singer, vol. 4, pp. 596–628. Oxford: University Press.

Explorers' Maps: Chapters in the Cartographic Record of Geographical Discovery. London: Routledge and Kegan Paul; New York: Praeger. (Reprint, with revisions, of a series of fourteen articles written for the *Geographical Magazine*, 1953–56.)

(With Gerald R. Crone and David B. Quinn) *Guide for Editors of the Hakluyt Society's Publications.* Cambridge: The Hakluyt Society.

"The Ordnance Survey, 1791–1825." *British Museum Quarterly* 21:59–61.

Review of *Captain Cook and Hawaii*, by David Samwell. *Geographical Journal* 124:562.

Review of *Surveys of the Seas*, by Mary Blewitt. *Mariner's Mirror* 44:172–74.

1959

"Leo Bagrow, Historian of Cartography and Founder of Imago Mundi, 1881–1957." *Imago Mundi* 14:5–12.

"A Medieval Map of Britain." Review of *The Map of Great Britain, circa A.D. 1360, Known as the Gough Map*, facsimile, with an introduction by E. J. S. Parsons. *Geographical Journal* 125:237–39.

Review of *The Southeast in Early Maps*, by William P. Cumming. *Geographical Journal* 125:263–64.

1960

Articles on Philippe Buache, Guillaume Delisle, Jules Sébastien César Dumont D'Urville, and Ruy Gonzalez de Clavijo. In *Encyclopaedia Britannica.* (These articles have appeared in all editions since 1960.)

"The Cartography of Columbus' First Voyage." in *The Journal of Christopher Columbus*, edited by L. A. Vigneras, pp. 217–27. London: Anthony Blond & The Orion Press. (Also issued as a Hakluyt Society Publication, Extra ser., no. 38.)

"Colour in Mapmaking." *Geographical Magazine* 32:544–53.

"Copyright and Piracy in Eighteenth-Century Chart Publication." *Mariner's Mirror* 46:207–12.

Discussion of "History in a Map," by E. M. Yates. *Geographical Journal* 126:49.

"Early Atlases." *Geographical Magazine* 32:529–43.
"An Elizabethan Naval Tract." *British Museum Quarterly* 22:51–53.
"Four English County Maps, 1602–3." *British Museum Quarterly* 22:47–50.
"The Map Room, British Museum." *Geographical Journal* 126:367–68.
(With H. G. Whitehead and P. D. A. Harvey) *Prince Henry the Navigator and Portuguese Maritime Enterprise . . . Catalogue of an Exhibition at the British Museum, September–October 1960.* London: Trustees of the British Museum.
Review of *Bibliographie der Gesamtkarten der Schweiz von Anfang bis 1802*, by Walter Blumer. *Imago Mundi* 15:123–24.
Review of *La Bibliothèque Nationale pendant les années 1952 à 1955, Rapport*, published by the Bibliothèque Nationale. *Journal of Documentation* 16:92–93.

1961

Articles on Affonso de Albuquerque, Francisco de Almeida, Louis Antoine de Bougainville, João de Castro, Gerardus Mercator, William Moorcroft, and George Vancouver. Also, section of the article "Map" entitled "History: Early Modern Period." In *Encyclopaedia Britannica.* (These articles have appeared in all editions since 1961 with the exception of that on William Moorcroft, which appeared from 1961 to 1965.)
"The Cartographic Record of the Discovery of North America: Some Problems and Paradoxes." In Congresso Internacional de História dos Descobrimentos, *Actas*, vol. 2, pp. 343–63. Lisbon: Comissão Executiva das Comemorações do V. Centenário da Morte do Infante D. Henrique.
"The Cartography of the Voyages." in *The Cabot*

Voyages and Bristol Discovery Under Henry VII, edited by J. A. Williamson, pp. 295–325. Hakluyt Society Publications, 2d ser., vol. 120. Cambridge: The Hakluyt Society.

Discussion of "The Ordnance Survey and Archaeology, 1791–1960," by C. W. Phillips. *Geographical Journal* 127:8–9.

"English Knowledge of the Portuguese Discoveries in the Fifteenth Century: A New Document." In Congresso Internacional de História dos Descobrimentos, *Actas*, vol. 2, pp. 365–74. Lisbon: Comissão Executiva das Comemorações do V Centenário da Morte do Infante D. Henrique.

"Map Room of the British Museum: Recent Acquisitions." *Geographical Journal* 127:257–58.

"Maps and the Local Historian," Middlesex Local History Council, *Bulletin* 11:11–21.

Review of *Philippine Cartography (1320–1899)*, by Carlos Quirino. *Geographical Journal* 127:125–26.

Review of *Der Stille Ozean: Entdeckung und Erschliessung*, by Hans Plischke. *Geographical Journal* 127:110.

1962

"The Charting of Australia by Flinders." In *Drawings by William Westall, Landscape Artist on Board H.M.S. Investigator during the Circumnavigation of Australia by Captain Matthews Flinders R.N., in 1801–1803*, edited by T. M. Perry and D. H. Simpson, pp. 29–32. London: Royal Commonwealth Society.

"Comment on Gastaldi's Map of Africa." In *The Somali Peninsula: A New Light on Imperial Matters*, rev. ed., appendix 20, pp. 82–83. Mogadiscio: Somali Republic, Information Services.

"Mercator and English Geography in the Sixteenth Century." *Duisburger Forschungen* 6:158–70.

"The Origins of the Ordnance Survey of Great Britain." *Geographical Journal* 128:415–26, discussion, 429.
"The Royal Map Collections." *British Museum Quarterly* 26:1–6.

1963

Bibliographical note in *Claudius Ptolemaeus: Cosmographia (Bologna, 1477)*, pp. v–xii. Theatrum Orbis Terrarum, 1st ser., vol. 1. Amsterdam: Theatrum Orbis Terrarum.
Bibliographical note in *Claudius Ptolemaeus: Cosmographia (Ulm, 1482)*, pp. v–xi. Theatrum Orbis Terrarum, 1st ser., vol. 2. Amsterdam: Theatrum Orbis Terrarum.
"Edzer Roukema, 1895–1960." *Imago Mundi* 17:6.
"An Eminent Geography Teacher." Review of *The History of Geography*, by J. N. L. Baker. *Geographical Journal* 129:508–09.
(With Leo Bagrow) *Meister der Kartographie*. Translated by Hermann Thiemke. Berlin: Safari-Verlag. (A translation of the Bagrow-Skelton *History of Cartography*, first published in English in 1964.)
"Philippine Cartography in the British Museum." in *Philippine Cartography (1320–1899)*, 2d rev. ed., by Carlos Quirino. Amsterdam: N. Israel.
"Ralegh as a Geographer." *Virginia Magazine of History and Biography* 71:131–49.
Review of *Ireland in Maps: An Introduction*, by John Andrews. *The Library*, 5th ser., 18:77.

1964

Article on Matthew Flinders. In *Encyclopaedia Britannica*. (This article has appeared in all editions since 1964.)
Bibliographical note in *Abraham Ortelius: Theatrum orbis terrarum (Antwerp, 1570)*, pp. v–xi. Theatrum Orbis

Terrarum, 1st ser., vol. 3. Amsterdam: Theatrum Orbis Terrarum.

Bibliographical note in *Cornelius à Wytfliet: Descriptionis Ptolemaicae augmentum (Louvain, 1597)*, pp. v–xxi. Theatrum Orbis Terrarum, 1st ser., vol. 5. Amsterdam: Theatrum Orbis Terrarum.

Bibliographical note in *Lucas Jansz. Waghenaer: Spieghel der zeevaerdt (Leyden, 1584–85)*, pp. v–xi. Theatrum Orbis Terrarum, 1st ser., vol. 4. Amsterdam: Theatrum Orbis Terrarum.

Bibliographical note in *Willem Jansz. Blaeu: The Light of Navigation (Amsterdam, 1612)*, pp. v–xiii. Theatrum Orbis Terrarum, 1st ser., vol. 6. Amsterdam: Theatrum Orbis Terrarum.

Contributions to *Mappemondes, A.D. 1200–1500*, edited by Marcel Destombes, pp. xv–xx, 149–53, 202–3, 214–21, 229–34, 239–41, and (with M. Destombes) 243–54. Monumenta cartographica vetustioris aevi, vol. 1; Imago Mundi Supplements, vol. 4. Amsterdam: N. Israel.

County Atlases of the British Isles, 1579–1850: A Bibliography. Part 1. Map Collectors' Series, no. 9. London: Map Collectors' Circle.

County Atlases of the British Isles, 1579–1850: A Bibliography. Part 2 (1612–1646). Map Collectors' Series, no. 14. London: Map Collectors' Circle.

"The Early Map Printer and His Problems." *Penrose Annual* 57:170–86.

The European Image and Mapping of America, A.D. 1000–1600. The James Ford Bell Lectures, no. 1. Minneapolis: Associates of the James Ford Bell Collection.

(With Leo Bagrow) *History of Cartography*. London: C. A. Watts; Cambridge: Harvard University Press. (A revised and augmented version of the 1951 German edition of Bagrow's *Die Geschichte der Kartographie*, as translated by D. L. Paisey.)

"John White's Contribution to Cartography." In *The American Drawings of John White, 1577–1590, with Drawings of European and Oriental Subjects,* edited by Paul H. Hulton and David B. Quinn, vol. 1, pp. 52–57. London: Trustees of the British Museum.
"Leo Bagrow and His *History of Cartography.*" *Bookseller,* 3 October 1964, pp. 1626–28.
Letter, in reply to E. G. R. Taylor's review of *Imago Mundi,* vol. 16. *Geographical Journal* 130:193–94.
"A Portuguese Sea-atlas of the Sixteenth Century." *Imago Mundi* 18:89.

1965

Bibliographical note in *Gerard de Jode: Speculum orbis terrarum (Antwerp, 1578),* pp. v–xiv. Theatrum Orbis Terrarum, 2d ser., vol. 2. Amsterdam: Theatrum Orbis Terrarum.
Bibliographical note in *Livio Sanuto: Geographia . . . dell' Africa (Venice, 1588),* pp. v–x. Theatrum Orbis Terrarum, 2d ser., vol. 1. Amsterdam: Theatrum Orbis Terrarum.
Bibliographical note in *Lucas Jansz. Waghenaer: Thresoor der zeevaert (Leyden, 1592),* pp. v–xv. Theatrum Orbis Terrarum, 2d ser., vol. 2. Amsterdam: Theatrum Orbis Terrarum.
Decorative Printed Maps of the Fifteenth to Eighteenth Centuries: A Revised Edition of "Old Decorative Maps and Charts," by A. L. Humphreys. London: Spring Books. (Reprint.)
"Did Columbus or Cabot See the Map?" *American Heritage* 16, no. 6, pp. 103–6. (Excerpt from *The Vinland Map and the Tartar Relation.*)
Discussion of "Lucas Janszoon Waghenaer: A Sixteenth Century Marine Cartographer," by C. Koeman. *Geographical Journal* 131:213–14.

Discussion of "New Light on the Hereford Map," by G. R. Crone. *Geographical Journal* 131:459, 462.
"Further Investigation." *Times* (London), 20 October 1965, p. 13. (Reply to an article by G. R. Crone.)
Introduction to *Braun and Hogenberg: Civitates orbis terrarum (Cologne and Antwerp, 1572–1618)*, pp. vii–xxiii. Mirror of the World, 1st ser., vol. 1. Amsterdam: Theatrum Orbis Terrarum.
Introductory essay in *James Cook, Surveyor of Newfoundland, Being a Collection of Charts of the Coasts of Newfoundland and Labrador, etc.*, pp. 5–32. San Francisco: David Magee.
"John Norden's View of London, 1600: A Study of the View." *London Topographical Record* 22:14–25.
Looking at an Early Map. University of Kansas Publications, Library Series, 17. Lawrence, Kansas: University of Kansas Libraries.
"The Mapping of Vinland." *American Heritage* 16, no. 6:9–10, 99–100. (Excerpt from *The Vinland Map and the Tartar Relation.*)
(Editor, with David B. Quinn) *The Principall Navigations, Voiages & Discoveries of the English Nation*, by Richard Hakluyt. Hakluyt Society Publications, Extra ser., no. 39. Cambridge: University Press, for the Hakluyt Society and the Peabody Museum of Salem.
Review of *Die alten Städtebilder*, by Friedrich Bachmann. *Geographical Journal* 131:566.
Review of *Carta Marina: World Geography in Strassburg, 1525*, by Hildegard B. Johnson. *Geographical Review* 55:307–8.
Review of *Catalogue des cartes nautiques sur vélin conservées au Département des Cartes et Plans*, published by the Bibliothèque Nationale. *Mariner's Mirror* 51:87.
Review of *Lucas Janszoon Waghenaer: De Spiegel der Zeevaerdt, 1584/1585*, facsimile ed., with an introduction by C. Koeman. *Mariner's Mirror* 51:280–82.

(With Thomas E. Marston and George D. Painter) *The Vinland Map and the Tartar Relation.* New Haven and London: Yale University Press.

"Was There a Lasting Colony?" *American Heritage* 16, no. 6:100–101. (Excerpt from *The Vinland Map and the Tartar Relation.*)

1966

Article on Giovanni Battista Ramusio. In *Encyclopaedia Britannica.* (This article has appeared in all editions since 1966.)

Articles on John and Sebastian Cabot. In *Dictionary of Canadian Biography,* vol. 1, pp. 146–58. Toronto: University of Toronto Press.

Bibliographical note in *Benedetto Bordone: Libro . . . de tutte l'isole del mondo (Venice, 1528),* pp. v–xii. Theatrum Orbis Terrarum, 3d ser., vol. 1. Amsterdam: Theatrum Orbis Terrarum.

Bibliographical note in *Claudius Ptolemaeus: Cosmographia (Rome, 1478),* pp. v–xiii. Theatrum Orbis Terrarum, 2d ser., vol. 6. Amsterdam: Theatrum Orbis Terrarum.

Bibliographical note in *Claudius Ptolemaeus: Cosmographia (Strassburg, 1513),* pp. v–xxii. Theatrum Orbis Terrarum, 2d ser., vol. 4. Amsterdam: Theatrum Orbis Terrarum.

Bibliographical note in *Claudius Ptolemaeus: Geographia (ed. Sebastian Münster, Basle, 1540),* pp. v–xxiii. Theatrum Orbis Terrarum, 3d ser., vol. 5. Amsterdam: Theatrum Orbis Terrarum.

Bibliographical note in *Francesco Berlinghieri: Geographia (Florence, 1482),* pp. v–xiii. Theatrum Orbis Terrarum, 3d ser., vol. 4. Amsterdam: Theatrum Orbis Terrarum.

Bibliographical note in *John Speed: A Prospect of the Most Famous Parts of the World (London, 1627),* pp. v–xv. Theatrum Orbis Terrarum, 3d ser., vol. 6. Amsterdam: Theatrum Orbis Terrarum.

Bibliographical note in *Lucas Jansz. Waghenaer: The Mariners Mirrour (London, 1588)*, pp. v–xi. Theatrum Orbis Terrarum, 3d ser., vol. 2. Amsterdam: Theatrum Orbis Terrarum.

Introduction to *Carolus Allard: Orbis habitabilis oppida et vestitus (Amsterdam, ca. 1695)*, pp. v–viii. Mirror of the World, 1st ser., vol. 4. Amsterdam: Theatrum Orbis Terrarum.

Letters, commenting on G. R. Crone's review of *The Vinland Map and the Tartar Relation. Geographical Journal* 132:177–78, 336–39, 448–50.

"The Vinland Map." *Journal of the Institute of Navigation* 19:271. (Reply to criticisms voiced by E. G. R. Taylor.)

"The Vinland Map and the Tartar Relation." *Geographical Magazine* 38:662–68.

1967

"Exhibition of the Mapping of Britain, Thirteenth–Nineteenth Centuries." In Twentieth International Geographical Congress, *Proceedings*, p. 199. London: Nelson, for the 20th International Geographical Congress.

"Historical Notes on Imago Mundi." *Imago Mundi* 21: 109–10.

"The Legend of a Welsh Columbus." Review of *Madoc and the Discovery of America*, by Richard Deacon. *Geographical Magazine* 40:65.

"Map Compilation, Production and Research in Relation to Geographical Exploration." In *The Pacific Basin: A History of Its Geographical Exploration*, edited by Herman R. Friis, pp. 40–56. AGS Special Publications, no. 38. New York: American Geographical Society.

(With R. V. Tooley) *The Marine Surveys of James Cook in North America 1758–1768, Particularly the Survey of Newfoundland: A Bibliography of Printed Charts and Sailing-Directions*. Map Collectors' Series, no. 37. London: Map Collectors' Circle.

"The Military Survey of Scotland, 1747–1755." *Scottish Geographical Magazine* 83:5–16. (Also issued separately as the Royal Scottish Geographical Society's Special Publication no. 1.)

Review of *Abraham Ortelius: Theatrum Orbis Terrarum, 1570*, facsimile ed., with an introduction by C. Koeman. *Imago Mundi* 21:123.

Review of *Kartenmacher aller Länder und Zeiten*, by Wilhelm Bonacker. *Imago Mundi* 21:125.

"Rodolfo Gallo, 1881–1964." *Imago Mundi* 21:115.

"The Vinland Map." *Journal of the Society of Archivists* 3:221–29.

1968

Bibliographical note in *Abraham Ortelius: The Theatre of the Whole World (London, 1606)*, pp. v–xxii. Theatrum Orbis Terrarum, 4th ser., vol. 4. Amsterdam: Theatrum Orbis Terrarum.

Bibliographical note in *Mercator-Hondius-Janssonius: Atlas or A Geographicke Description of the Regions, Countries and Kingdomes of the World (Amsterdam, 1636)*, pp. v–xxvii. Theatrum Orbis Terrarum, 4th ser., vols. 2 and 3. Amsterdam: Theatrum Orbis Terrarum.

"A Contract for World Maps at Barcelona, 1399–1400." *Imago Mundi* 22:107–13.

"Cook's Contribution to Marine Survey," *Endeavour* 27: 28–32.

County Atlases of the British Isles, 1579–1850: A Bibliography. Part 3 (1627–1670). Map Collectors' Series, no. 41. London: Map Collectors' Circle.

County Atlases of the British Isles, 1579–1850: A Bibliography. Part 4 (1671–1703). Map Collectors' Series, no. 49. London: Map Collectors' Circle.

Discussion of "Adolf Erik Nordenskiöld (1832–1901): Polar Explorer and Historian of Cartography," by George Kish. *Geographical Journal* 134:502.

"The First English World Atlases." In *Kartengeschichte und Kartenbearbeitung: Festschrift zum 80. Geburtstag von Wilhelm Bonacker*, edited by Karl-Heinz Meine, pp. 77–81. Bad Godesberg: Kirschbaum Verlag.

"First International Meeting on Nautical History." *Imago Mundi* 22:117.

Letter, requesting information on the history of the Ordnance Survey, to be used in writing the official history. *Geographical Journal* 134:308.

(With D. G. Moir) "New Light on the First Atlas of Scotland." *Scottish Geographical Magazine* 84:149–59.

"Obituary, Professor E. G. R. Taylor." *Imago Mundi* 22: 114–15.

Review of *Goa e as Praças do Norte* and *Imagens de Macau*, by Raquel Soeiro de Brito. *Geographical Journal* 134:578–79.

Review of *Maps and Plans in the Public Record Office. I: British Isles, c. 1410–1860. Imago Mundi* 22:120–21.

1969

Bibliographical note in *Claudius Ptolemaeus: Geographia (Venice, 1511)*, pp. v–xi. Theatrum Orbis Terrarum, 5th ser., vol. 1. Amsterdam: Theatrum Orbis Terrarum.

Captain James Cook—After Two Hundred Years: A Commemorative Address Delivered Before the Hakluyt Society. London: Trustees of the British Museum.

(With G. de Boer) "The Earliest English Chart with Soundings." *Imago Mundi* 23:9–16.

Foreword to *Mercator: A Monograph on the Lettering of Maps, etc., in the Sixteenth Century Netherlands with a Facsimile and Translation of His Treatise on the Italic Hand and a Translation of Ghim's "Vita Mercatoris,"* by A. S. Osley, pp. 9–10. London: Faber and Faber; New York: Watson Guptill.

(With P. D. A. Harvey) "Local Maps and Plans Before 1500." *Journal of the Society of Archivists* 3:496–97.

"Magellan's Voyage," *Geographical Magazine* 42:219–23.

(Editor and translator) *Magellan's Voyage: A Narrative Account of the First Circumnavigation*, by Antonio Pigafetta. 2 vols. New Haven: Yale University Press.

(With P. D. A. Harvey) "Medieval English Maps and Plans." *Imago Mundi* 23:101–2.

Review of *American Activities in the Central Pacific, 1790–1870*, edited by R. Gerard Ward. *Mariner's Mirror* 55: 108–9.

Review of *A History of Cartography: 2,500 Years of Maps and Mapmakers*, by R. V. Tooley, Charles Bricker, and G. R. Crone. *Imago Mundi* 23:116.

Review of *Itineraria picta: Contributo allo studio della Tabula Peutingeriana*, by Annalina and Mario Levi. *Imago Mundi* 23:115–16.

Review of *Maps and Charts Published in America before 1800: A Bibliography*, by J. C. Wheat and C. F. Brun. *Imago Mundi* 23:116–17.

Review of *The Mathematical Practitioners of Hanoverian England, 1714–1840*, by E. G. R. Taylor. *Mariner's Mirror* 55:109–10.

A Venetian Terrestrial Globe, Represented by the Largest Surviving Printed Gores of the Sixteenth Century. Bologna: Garisenda Antiquariato.

1970

(With Coolie Verner) Bibliographical note in *John Thornton: The English Pilot, the Third Book (London, 1703)*, pp. v–xv. Theatrum Orbis Terrarum, 5th ser., vol. 3. Amsterdam: Theatrum Orbis Terrarum.

(Editor) *Charts and Views Drawn by Cook and His Offi-*

cers and Reproduced from the Original Manuscripts. Cambridge: For the Hakluyt Society at the University Press. (Reprint of the 1955 edition.)

"Cook's Voyages." *Commonwealth: The Journal of the Royal Commonwealth Society* 14:197–201.

County Atlases of the British Isles, 1579–1850: A Bibliography. Part 5 (1579–1703: appendixes). Map Collectors' Series, no. 63. London: Map Collectors' Circle.

County Atlases of the British Isles, 1579–1850: A Bibliography. [Vol. 1], *1579–1703.* London: Carta Press. (Republication, with additions, of material originally published in the Map Collectors' Series.)

Explorers' Maps: Chapters in the Cartographical Record of Geographical Discovery. Feltham, N. Y.: Spring Books. (Reprint.)

Foreword to *Alexander Dalrymple (1737–1808) and the Expansion of British Trade,* by Howard T. Fry, pp. xiii–xiv. Toronto: University of Toronto Press for the Royal Commonwealth Society.

"The Influence of Verrazzano on Sixteenth Century Cartography." In *Giovanni da Verrazzano: Giornate commemorative, Firenze, Greve in Chianti, 21–22 ottobre 1961,* pp. 55–69. Florence: L. S. Olschki.

Introduction to *Giovanni Battista Ramusio: Navigationi et viaggi (Venice, 1563–1606),* pp. v–xvi. Mundus Novus, 1st ser., vols. 2–4. Amsterdam: Theatrum Orbis Terrarum.

Introduction to *Two Hundred and Fifty Years of Mapmaking in the County of Sussex,* edited by Harry Margary, p. 1. Lympne Castle, Kent: Harry Margary; Chichester, Sussex: Phillimore & Co.

"The Military Surveyor's Contribution to British Cartography in the Sixteenth Century." *Imago Mundi* 24: 77–83.

Remarks on being awarded the Victoria Medal of the

Royal Geographical Society. *Geographical Journal* 136:506.

Review of *Maps and Charts Published in America Before 1800: A Bibliography*, by J. C. Wheat and C. F. Brun. *The Library* 5th ser. 25:272–74.

Review of *Northern Mists*, by Carl O. Sauer. *Mariner's Mirror* 56:251.

The Seaman and the Printer. Publicações do Agrupamento de Estudos de Cartografia Antiga, série de separatas, 41. Coimbra: Junta de Investigações do Ultramar (Secção de Coimbra).

1971

(With John Summerson) *A Description of Maps and Architectural Drawings in the Collection Made by William Cecil, First Baron Burghley, Now at Hatfield House*. Oxford: Roxburghe Club.

(With William P. Cumming and David B. Quinn) *The Discovery of North America*. London: Elek Books.

Discussion of papers in *Proceedings of the Vinland Map Conference*, edited by Wilcomb E. Washburn. Studies in the History of Discoveries. Chicago: For the Newberry Library by the University of Chicago Press.

1972

"Maps in the Society's Library—II," Royal Commonwealth Society, *Library Notes* n.s. 177:1–3.

In Press or Forthcoming

"Hakluyt's Maps." In *A Handbook to Hakluyt*, edited by David B. Quinn. Cambridge: For the Hakluyt Society at the University Press.

"The Le Moyne–De Bry Map." In *Jacques Le Moyne de Morgues: A Huguenot Artist in France, Florida and England*, edited by Paul H. Hulton. London: British Museum.

(With Marcel Destombes) *Monumenta cartographica vetustioris aevi*, vol. 4.

(With A. D. Baxter and S. T. M. Newman) *Saxton's Survey of England and Wales and the Maps from It; With a Facsimile of Saxton's Wall-Map of 1583.* Imago Mundi Supplements, 6. Amsterdam: N. Israel.

(With Jeannette D. Black) "Too Many Cooks." (An aid to the identification of three officers in the Royal Navy named James Cook, all of whom were making charts in American waters in the 1760s.)

Index

Index

Index